B. Kienzler, C. Borkel, V. Metz and M. Schlieker

Long-Term Interactions of Full-Scale Cemented Waste Simulates with Salt Brines

Karlsruhe Institute of Technology
KIT SCIENTIFIC REPORTS 7721

Long-Term Interactions of Full-Scale Cemented Waste Simulates with Salt Brines

by

B. Kienzler, C. Borkel, V. Metz and M. Schlieker

KIT Scientific Publishing

Report-Nr. KIT-SR 7721

Impressum

KIT Scientific Publishing

Karlsruher Institut für Technologie (KIT)
KIT Scientific Publishing
Straße am Forum 2
D-76131 Karlsruhe

KIT Scientific Publishing is a registered trademark of Karlsruhe Institute of Technology. Reprint using the book cover is not allowed.

www.ksp.kit.edu

Print on Demand 2016

ISSN 1869-9669
ISBN 978-3-7315-0576-1
DOI: 10.5445/KSP/1000058276

Abstract

Since 1967 radioactive wastes have been disposed of in the Asse II salt mine in Northern Germany. A significant part of these wastes originated from the pilot reprocessing plant WAK in Karlsruhe and consisted of cemented $NaNO_3$ solutions bearing fission products, actinides, as well as process chemicals. With respect to the long-term behavior of these wastes, the licensing authorities requested leaching experiments with full scale samples in relevant salt solutions which were performed since 1979. The experiments aimed at demonstrating the transferability of results obtained with laboratory samples to real waste forms and at the investigation of the effects of the industrial cementation process on the properties of the waste forms. This research program lasted until 2013. The corroding salt solutions were sampled several times and after termination of the experiments, the solid materials were analyzed by various methods. The results presented in this report cover the evolution of the solutions and the chemical and mineralogical characterization of the solids including radionuclides and waste components, and the paragenesis of solid phases (corrosion products). The outcome is compared to the results of model calculations. For safety analysis, conclusions are drawn on radionuclide retention, evolution of the geochemical environment, evolution of the density of solutions, and effects of temperature and porosity of the cement waste simulates on cesium mobilization.

Zusammenfassung

Seit 1967 wurden radioaktive Abfälle in der Schachtanlage Asse II in Norddeutschland eingelagert. Ein wesentlicher Teil dieser Abfälle stammte aus der Pilot-Wiederaufarbeitungsanlage WAK in Karlsruhe und bestand aus den sauren $NaNO_3$-haltigen Verdampferkonzentraten welche Spaltprodukte, Actinide sowie Prozesschemikalien enthielten. Diese Abfälle wurden damals weitgehend zementiert. In Bezug auf das Langzeitverhalten dieser Abfälle forderten die Genehmigungsbehörden Auslaug- und Korrosionsexperimente mit Abfallgebinden im realen Maßstab in den relevanten Salzlösungen. Diese Experimente wurden seit 1979 durchgeführt, um die Übertragbarkeit von Ergebnissen aus Experimenten mit Laborproben auf echte Abfallprodukte nachzuweisen sowie die Eigenschaften der in einem industriellen Zementierungsprozess hergestellten Produkte zu untersuchen. Dieses Forschungsprogramm dauerte bis 2013 an. Die Salzlösungen wurden mehrmals beprobt und nach Beendigung der Versuche wurden die Feststoffe mit verschiedenen Methoden untersucht. In diesem Bericht sind die Entwicklung der korrodierenden Lösungen sowie die chemische und mineralogische Charakterisierung der Feststoffe einschließlich der Radionuklide und weiterer Abfallbestandteile dokumentiert, und die Paragenese der Korrosionsprodukte beschrieben. Die Befunde werden mit den Ergebnissen von Modellrechnungen verglichen. Für Sicherheitsanalysen werden Schlussfolgerungen bezüglich der Radionuklidrückhaltung, der Entwicklung des geochemischen Milieus und der Entwicklung der Dichten der Lösungen gezogen. Der Einfluss der Temperatur und der Porosität der Zementprodukte bezüglich der Mobilisierung von Cäsium wird dargestellt.

Table of Contents

List of Figures

List of Tables

List of Abbreviations

ANKA	Synchrotron Radiation Facility at the Karlsruhe Institute of Technology
BSE	Backscattered electrons
CEA	Commissariat à l'énergie atomique et aux énergies alternatives
C-S-H	calcium silicate hydrate
EXAFS	extended X-ray absorption fine structure
FZJ	Research Center Jülich
GSF	Gesellschaft für Strahlen- und Umweltforschung
HDB	Hauptabteilung Dekontaminationsbetriebe, Waste Management Facility of the WAK GmbH
HOZ	Slag cement (HOZ 35 L, present notation CEM III/A 32,5 N)
HR-ICP-MS	High resolution inductively coupled plasma - mass spectrometry
ICP-MS	Inductively coupled plasma - mass spectrometry
IfT	Institut für Tieflagerung in the GSF
INE	Institut für Nukleare Entsorgung Institute for Nuclear Waste Disposal
KfK	Kernforschungszentrum Karlsruhe, Research Center Karlsruhe
KIT	Karlsruhe Institute of Technology
m_{H+}	proton concentration in solution in molal units
PZ	(Ordinary) Portland cement (PZ 35 F, PZ 45 HS, present notation CEM I 32.5 R)
SEM-EDX	scanning electron microscopy with energy dispersive X-ray detector
W/C ratio	water (volume) to (mass of) cement ratio, unit: l/kg
XANES	X-ray absorption near edge structure
XRD	X-ray powder diffractometry
WAK	Pilot reprocessing plant at the Research Center Karlsruhe
γ	chemical activity coefficient

1 Introduction

Non-heat-producing radioactive wastes (low and intermediate level radioactive wastes) accrued as solid or liquid primary wastes during operation of nuclear power plants or from decommissioning of nuclear facilities. For safe disposal, these wastes are processed: Burnable wastes are incinerated and solids or ashes are compacted. Sludges, evaporator- and filter-concentrates as well as ion exchangers are dried before stabilization and solidification by mixing with cement powder and water, resulting in a solid waste form. The cementation process is a widely used and proven procedure. Benefits of cementation are the relatively simple process technique, applicable to different waste streams and low cost of the matrix material. Disadvantages are the relatively low waste load in the final product and potential chemical interaction between cement and some waste components. For example, waste solutions containing boric acid (i.e. evaporator concentrates from pressure water reactors) or ion exchanger resins may react with the cement matrix and prevent setting and hardening. During the operation of the pilot reprocessing plant WAK at Karlsruhe, low and intermediate level radioactive evaporator concentrates were solidified by cementation.

1.1 Background

In 1978, before termination of radioactive waste disposal activities in the Asse II salt mine, the waste treatment facility (HDB) of the research center Karlsruhe delivered 13299 waste containers to the Asse II salt mine [1]. Of these containers 4074 contained cemented radioactive evaporator concentrates from the reprocessing plant WAK. In total, 5181 containers with cemented wastes were produced and transferred to Asse II in this year. The average α activity amounted to about 3.7 GBq, the average β/γ activity to 226 GBq. The activity of single containers showed significant deviations from the average. Due to the activity load especially by actinides of the cemented evaporator concentrates from the WAK, these wastes gained importance in the scope of safety analysis.

Resistance against leaching was considered as main criterion for evaluating the quality of solidified radioactive wastes. Important properties of cemented waste forms are controlled by the production process. The correlation between production technique and leaching behavior got high priority in research organizations and by the licensing authorities. Determination of the properties of concrete or cemented waste forms were requirements of the license for disposal of low level radioactive wastes in the Asse II mine [2] with focus on the impact of water or brine on radionuclide mobility.

At the instigation of a review board on 'Treatment and Disposal of Radioactive Wastes', established by the Federal Ministry of Research and Technology, proposals dealing with the characterization and related control procedures for waste forms were elaborated [3]. In 1978, a workshop was organized by the Nuclear Research Center Karlsruhe on 'Cement and Concrete properties under Conditions Relevant for Disposal in the Asse II salt mine'. The workshop aimed at the 'Weighting and Completion of corrosion relevant properties of cement and concrete in salt brines', with emphasis on the type of cement, the water to cement-(W/C)ratio, the degree of consolidation, the hydration period, the plasticizer and other additives (sand, clay, bentonite, etc.) as well as waste load and sealing agents ('Zement Dichtungsmittel'). It was intended to derive recommendations for well suited types of cement, additives and cementation techniques.

Participants came from the licensing organization, cement and drilling companies, universities and research centers. Presentations covered among others the following topics:

- Literature review on cement corrosion in salt brines
- Corrosion sensitive properties and correlation with the strength of concrete
- Flash tests for cements
- Properties of sulfate resistant cements in NaCl-, $MgCl_2$-, $MgSO_4$-rich solutions
- Effect of NaCl or bentonite additives on properties of well cements ('Tiefbohrzement")
- Experiences with grouting, especially with magnesia cement in potash mines

The result of the workshop was the definition of a comprehensive research program on leaching and corrosion experiments. In this context, it was discussed if the investigated properties of laboratory samples can be extrapolated to real waste forms and how the technical cementation process affects the properties, especially the leachability of radionuclides. In the case of cemented radioactive wastes the pore structure affects the leaching and corrosion behavior and consequently the radionuclide retention [4, 5]. To test the transferability of mechanisms and properties that were scrutinized by the use of laboratory scale samples to real waste forms, and to investigate the influence of the technical cementation processes, leaching experiments were started in 1978 using full-scale cemented waste simulates at the Asse II mine. The simulates had the typical composition of waste forms which arose at the WAK from the PUREX process. The composition and properties of the waste concentrates are published [5]. The main component in the concentrate was sodium nitrate. The $NaNO_3$ load of the simulates was around 10 wt.-% of the cemented waste product. For investigation of the leaching behavior the simulates were doped with single radionuclides such as cesium, neptunium or uranium.

Throughout this report, the full-scale cemented waste simulates are referred to by cement blocks or cemented simulates.

1.2 Overview on cement related investigations

Cement is an inorganic construction material belonging to the so called binding agents. It hardens by chemical reactions with water (hydration) and forms a solid, hardened cement paste. Cement consists of about 58 - 66 wt.-% calcium oxide (CaO), 18 - 26 wt.-% silicon dioxide (SiO_2), 4 - 10 wt.-% aluminum oxide (Al_2O_3) and 2 - 5 wt.-% iron oxide (Fe_2O_3) [6]. During the calcination of the natural raw materials, limestone (mainly $CaCO_3$) and clay (or marl), clinker phases are formed in the kiln:

Tri-calcium silicate (Alit),	C_3S (3CaO SiO_2)
Di-calcium silicate (Belit),	C_2S (2CaO SiO_2)
Tri-calcium aluminate,	C_3A (3CaO Al_2O_3)
Tetra-calcium aluminate ferrite,	C_4AF bzw. $C_2(A,F)$ (4 CaO Al_2O_3 Fe_2O_3)

These are mixed with calcium sulfate and eventually grinding aids during grinding. When this cement is mixed with water the following main phases are formed during setting and hardening:

Calcium silicate hydrate	C-S-H ($CaO-SiO_2-H_2O$ of variable stoichiometry)
Portlandite	CH (Ca(OH)2).

According to DIN EN197-1, 27 types of cement compositions are defined which contain granulated slag, fly ash, limestone or gypsum, e.g., in different dosage. Most of the experiments performed in the context of the Asse II mine were performed with pure Portland cement (PZ 35 F, PZ 45 HS, present notation CEM I 32.5 R or CEM I 42.5 R). In some cases, slag cement HOZ 35 L (CEM III/A 32,5 N) was included. The cement blocks (simulates) produced for the full-scale experiments were made from PZ 35 F (CEM I 32.5 R). The chemical composition of the used batches of cement powder is not available.

At the time when the experiments were designed, rock salt was considered as host rock for disposal of radioactive wastes in Germany. For this reason, most investigations on waste forms extended to the behavior of waste forms in rock salt or in contact with solutions relevant for salt deposits consisting of different salt minerals. To cover the most important compositions, three solutions were agreed upon [7]. For most of the leaching and corrosion experiments the so called Q-brine ($MgCl_2$-rich solution at the in invariant point Q in the $Na^+/K^+/Mg^{2+}/Cl^-/SO_4^{2-}/H_2O$ system, referred to as 'solution 1' [7]) and saturated, NaCl-rich solution in contact with anhydrite or polyhalite (referred to 'solution 3' [7]) were used. The compositions are listed in Tab. I. The solution called 'solution 2' contains calcium and was used mainly for investigations of vitrified high level wastes.

Tab. I Composition of saturated salt solutions according to the High Level Waste Review Panel [7].

	solution 1 (mol / kg H_2O)	**solution 2** (mol / kg H_2O)	**solution 3** (mol / kg H_2O)
Na^+	0.4	0.08	5.98
K^+	1.0	0.02	
Mg^{2+}	4.4	5.35	0.02
Ca^{2+}		0.31	0.02
Cl^-	9.8	11.42	5.98
SO_4^{2-}	0.2	0.001	0.04
pH	5.5	5.7	5.9
Density (g / ml)	1.30	1.33	1.20
at temperature (°C)	70	25	22
Ionic strength (mol / kg)	14.4	17.1	6.1

1.2.1 Laboratory experiments

Laboratory experiments were performed using prismatic samples with a dimension of 2×2×5 cm^3. The investigations covered leaching and corrosion tests and also the effect of different additives or application of various process techniques (e.g. vibration). Other studies optimized the waste load, compared different types of cement powder and the water to cement ratio. The mechanical properties (compression strength and E-module) were analyzed as function of the porosity and waste load. A compendium of the investigations and the results was published by Vejmelka [5].

1.2.2 Experiments with full-scale samples without canisters

In literature few experiments with full-scale samples on leaching and corrosion of cemented radioactive wastes and on radionuclide retention have been described [8-15]. Most of these experiments have been performed for short times. In the Research Center Karlsruhe (today KIT) tests were run with two real-scale blocks (KfK #5 and KfK #6) for about 12 years in $MgCl_2$-rich brine (solution 1) at 40 °C. The evolution of calcium-, magnesium- and sulfate-concentrations was measured as well as the cesium release.

Between 1975 and 1990, leaching and corrosion tests were conducted with real cemented waste forms in the Research Center Karlsruhe and at CEA, France. The cement blocks were produced at HDB. The W/C ratio was 0.63 l/kg (identification of the radioactive samples: AF2 and AF3). The samples had a mass of 281 kg, a density of 1.7 kg/dm^3, a total α activity of 1.14×10^9 Bq/sample and and a total β/γ activity to 8.50×10^9 Bq/sample [16]. Each of these two samples was exposed to 400 l $MgCl_2$ solution (Q-brine) for a period of 4.5 years.

CEA, France, investigated further full-scale cemented or bituminized waste forms, containing α, β and γ emitters, originating from solid and liquid wastes [16]. Some Russian publications deal with long-term behavior of cemented waste forms, having nitrate loads of more than 30-wt.-%. This waste load is significantly higher than that of the waste forms under investigation in Germany [17].

After complete hydration of the cemented simulates the blocks were stored in 400 l containers, which were filled with solution. These solutions encompassed tap water, saturated NaCl-rich and $MgCl_2$-rich solutions (Q-brine). The experiments took place at the 490 m level of the Asse II salt mine in a separate gallery (Fig. 1). The simulated cement blocks were doped with natural cesium or with radioactive ^{137}Cs, with uranium and with ^{237}Np. The preparation of the blocks, the regular sampling of the solutions and the obtained results have been documented (internal reports [18-22]) and published [23, 24]. In 1998 the sampling was performed by the Freie Universität Berlin (FUB). Later, sampling was conducted by the Research Center Karlsruhe in 2001 [25], 2003 [26], 2006 [27] and 2011 [28]. In 2006, 4 full-scale cement blocks were removed from the corroding solutions, and the composition of the solids was investigated [29, 30]. After termination of all experiments in 2013, the full-scale simulates were transferred to KIT-INE for final analyses. Details have not yet been published (unpublished reports [28, 31]).

Additional to the mentioned publications, details from these long-term full-scale experiments have been presented [32, 33]. Modeling exercises have been published on the behavior of cesium [34] and uranium [35, 36]. The results of the long-term tests have also been used for derivation of radionuclide source terms for the Asse II salt mine [37] and in the scope of the preliminary safety analysis of Gorleben (vSG) [38].

Fig. 1 Photo of the separated gallery for the full-scale experiments at the 490 m level of the Asse II salt mine.

1.2.3 Experiments with full-scale samples in a canister

Since 1980 cemented low and intermediate level radioactive evaporator concentrates were tested including the steel drums or inserted in concrete shielding containers. The tests aimed at the resistance of the entire waste form – canister system against corroding solutions. The inactive simulates were produced using the industrial in-drum cementation process. Portland cement and slag cement were used (W/C ratio 0.44 l/kg). The steel canisters were sealed by a double lid system. After hydration of the cement, two samples were stored in $MgCl_2$ brine. Another two simulated waste forms were additionally encased in concrete shielding canisters (VBA) (characteristics are reported in the annex). The behavior of these simulates was investigated in Q-brine at a temperature of 40° C. As indicators for the leaching, the samples have been doped with inactive cesium [39].

During the test period of several years, no significant release of Cs was observed from the simulates in closed canisters. Single measurements resulted in slight Cs concentrations, but these were traced back to contaminations of the outer canister material prior to the start of experiments (personal notes by W. Kluger, 1980). In the case of the simulates in concrete shielding, neither nitrate nor cesium could be detected in the corroding solutions after 10.9 years. Only the iron concentration of the $MgCl_2$-rich solution increased. Fig. 2a shows the cut through the concrete shielding canister as well at the incorporated waste drum. A horizontal drilling core was obtained, its location is shown in Fig. 2a, also.

Mineralogical investigations of the drilling core (Fig. 2b) showed significant degradation to a depth of about 2.5 cm from the outer surface. In this layer, precipitation of salts, gypsum, sylvite and Sorel phases were found. To a depth of 5 cm, the binding of grains was disturbed and the corrosion product Friedel's Salt was detected (3 CaO Al_2O_3 $CaCl_2 \cdot 10\ H_2O$). Sulfate originating from the Q-brine penetrated to a depth of about 3 cm, magnesium to about 5 cm. Chloride penetrated to about 11 cm into the concrete shielding. The mortar used for fixation of the concrete lid of the shielding container showed heavy corrosive attack (see Fig. 2a top) [40].

a.) Picture of a cut through a concrete shielding container including a canister (drum) with cemented waste simulate. Bottom left: Location of a drilling core

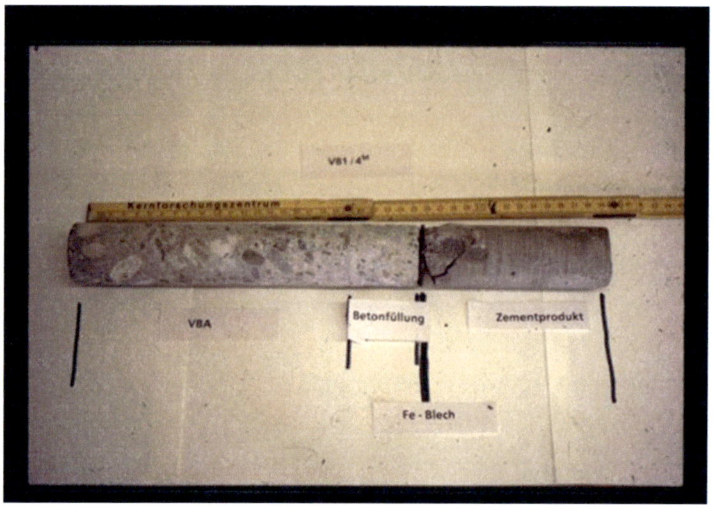

b.) Drilling core taken from a concrete shielding container including a canister (drum) with cemented waste simulate.

Fig. 2 Investigations on a full-scale simulate in presence of a concrete shielding container after 10.9 years in Q-brine at 40°C. Images show a cut through (top) and a drilling core (bottom) through a concrete shielding container.

2 Description of the full-scale leaching and corrosion experiments

2.1 Waste simulates

The leaching and corrosion experiments were performed with simulated, cemented evaporator concentrates of a composition according to the wastes from the WAK. The concentrates were analyzed and the properties have been published [5, 41]. The low and intermediate waste stream was highly relevant as it contained high β/γ activity as well as the highest α activity in comparison to other low and intermediate level waste streams. The main component of this waste stream was sodium nitrate, other metal nitrates, remaining tributylphosphate (TBP) and decontamination agents such as citrate, oxalate, ethylene diamine tetra acetic acid (EDTA) and detergents. Tables of the precise composition of this waste stream are available [29, 31].

2.2 Preparation of the full-scale simulates

The preparation of the full-scale simulates was described by Altenhain-Haese [42]. Different mixing techniques were applied for preparing the full-scale samples. The cement blocks #20 to #36 were fabricated at the Research Center Jülich (FZJ). The waste components were dissolved in 7.25 l 65% HNO_3 and adjusted to pH 9-10 using 45 % NaOH solution. For the blocks #31 to #36 86 l of dissolved waste simulate was prepared and mixed to the cement powder. For the industrial cementation process at HDB OPC type PZ 45 and in-drum mixing was applied [43]. The resulting cemented blocks had a $NaNO_3$ load of about 10 wt.-%. The cement blocks were hydrated at W/C ratios between 0.24 and 0.50 l/kg. The volumes of the cement blocks were produced in sizes from 83 l (∅ ~ 0.45 m, h ~ 0.60 m) to about 177 l (∅ ~ 0.57 m, h ~ 0.70 m). The theoretical densities of the blocks were in the range of 2.20 to 2.40 g/cm^3 for blocks #1 to #24 and between 2.02 and 2.14 g/cm^3 for blocks #25 to #36. After hydration of the simulated waste forms, the mould (Fig. 3) was removed. Some of the cement blocks developed rim zones which could be a result of insufficient mixing close to the wall of the mould. For handling the cement blocks (weight up to 360 kg), one or two steel holders were cemented into the blocks. Some minor problems were noticed during the production of the cement blocks:

- In the case of uranium doped block #34 the pH of the prepared simulate solution was above the intended value and it showed yellow color in contrast to the simultaneously produced solutions which were prepared for other uranium doped cement blocks (#31, #32, #33). Moreover, the mixer touched the mould; therefore the circulation radius of the mixer was reduced. This resulted in insufficiently mixed rim zones. The setting time of this block was found to be accelerated.
- In the case of block #31 segregation was observed. The segregation was explained by insufficient mixing time. In average the mixing of 200 l slurry added up to 8 minutes.
- The blocks #31 to #34 did not show visually inhomogeneous zones at the surfaces.
- During preparation of block #36 one package of cement powder was accidently spread to the floor. However, the cement was recollected and used.

Fig. 3 Photo of the mould used to prepare cement blocks #20 to #36.

Fig. 4 shows photos of some cement blocks after removal from the mould, before starting the leaching and corrosion tests.

#25 #28 #30

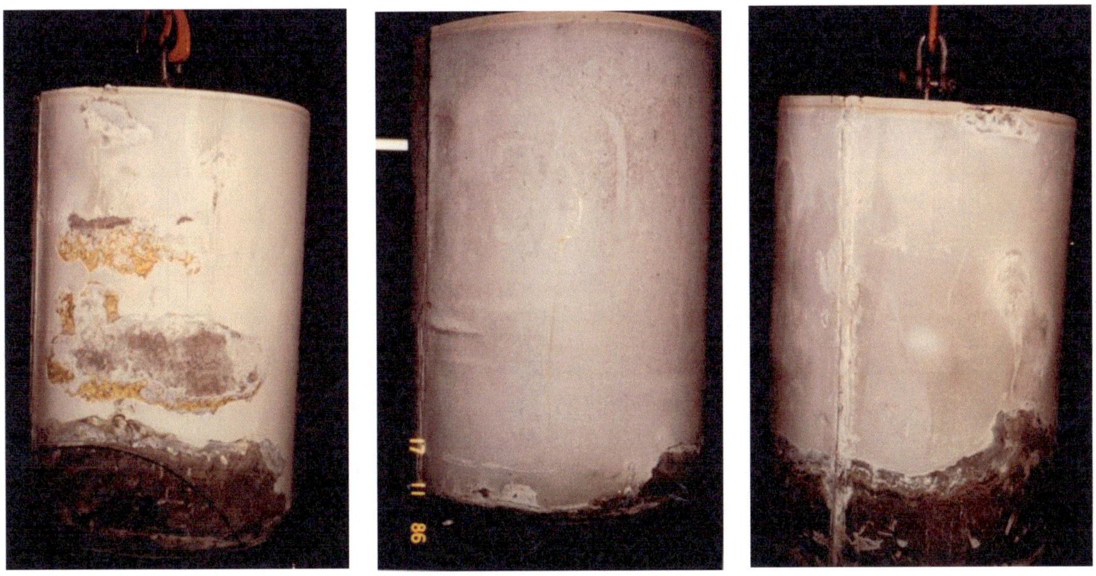

#33 #35 #36

Fig. 4 Photos of some cemented waste simulate blocks after removal from the mould, before starting the leaching and corrosion tests.

2.3 Test procedure

After setting and hardening of the cemented waste simulates the cement blocks were stored in 400 l vessels (equipped with glass fiber reinforced bisphenol resin liners), filled with the corroding solutions. The total volume of the leaching vessels was about 335 l. The experiments were kept at ambient temperature of the salt mine (28 $\pm$ 1 °C).

The inner diameter of the reinforced vessels was $\varnothing$ = 0.68 m and the total height h = 0.98 m and the full-scale cement blocks rested on pipe sockets of 0.05 m height in order to guarantee contact of the entire cement block surfaces to the solution. The lids of the 400 l leaching vessels were equipped with ducts for liquid sampling from different vertical levels of the solution.

Tab. II shows key characteristics of the full-scale cement waste simulate blocks conditions of the experiments, performed at the Asse II salt mine. The leachates tap water, NaCl solution and $MgCl_2$ solution (Q brine) were used.

Tab. II Key characteristics of the full-scale cemented simulate blocks and conditions of leach and corrosion experiments [42, 44, 45].

	Tracer	W/C ratio (l/kg)	Mass (kg)	Leachate	Volume of leachate (l)	Atmos-phere	Start of test	End of test
#1	none	0.4	470	Tap water	125 l +7.2 l in 06.03.1990		17.05.1979	2013
#2	none	0.4, 188 kg Sand	364	nat. NaCl-Solution	185 l		16.07.1979	2013
#4	none	0.29	249	nat. $MgCl_2$-Solution	175 l +11 l in 06.03.1990		20.08.1979	2013
#5	none	0.4	294	nat. NaCl-Solution	200 l		16.07.1979	2013
#6	none	0.4, 126 kg Sand	211	Tap water	213 l +18 l in 06.03.1990		09.07.1979	2013
#7	Cesium (223 g)	0.24	186	nat. $MgCl_2$-Solution	250 l		01.10.1979	2013
#8	Cesium (233 g)	0.24	194	nat. $MgCl_2$-Solution	238 l since 13.10.1983 187 l +7 l in 06.03.1990		01.10.1979	2013
#9	Cesium (236 g)	0.24	197	nat. $MgCl_2$-Solution	247 l since 13.10.1983 196 l +7 l in 06.03.1990		01.10.1979	before 2001
#10	Cesium (246 g)	0.24	205	nat. NaCl-Solution	235 l		01.10.1979	before 2001
#11	Cesium (253 g)	0.24	211	nat. NaCl-Solution	242 +6.0 l in 06.03.1990		01.10.1979	2013
#12	Cesium (253 g)	0.24	211	nat. NaCl-Solution	235 +3.5 l in 06.03.1990		01.10.1979	2013
#13	Cesium (231 g)	0.24	192	Tap water	239 +5.5 l in 06.03.1990		01.10.1979	2013
#14	Cesium (232 g)	0.24	193	Tap water	238 l +5.5 l in 06.03.1990		01.10.1979	2013
#15	Cesium (237 g)	0.24	197	Tap water	239 l +5.5 l in 06.03.1990		01.10.1979	2013
#20	Cesium (509 g)	0.32	199	nat. $MgCl_2$-Solution	243 l +7.0 l in 06.03.1990		11.11.1980	2013

	Tracer	W/C ratio (l/kg)	Mass (kg)	Leachate	Volume of leachate (l)	Atmos-phere	Start of test	End of test
#21	^{137}Cs (342 MBq reference date 27.10.1980)	0.32	205	nat. $MgCl_2$-Solution	246 l replaced in a different vessel in 03.06.1986 +7.2 l in 06.03.1990		02.06.1981	2013
#22	^{137}Cs (342 MBq reference date 27.10.1980))	0.32	205	nat. $MgCl_2$-Solution	246 l		02.06.1981	2013
#23	^{137}Cs (344 MBq reference date 27.10.1980))	0.32	205	nat. NaCl-Solution	246 l +11.0 l in 06.03.1990		07.07.1981	2013
#24	^{137}Cs (658 MBq reference date 27.10.1980))	0.32	205	nat. $MgCl_2$-Solution	246 l		07.07.1981	2013
#25	$Uranium_{nat.}$ (951 g)[#]	0.30	363	nat. $MgCl_2$-Solution	120 l +17.58 l until 12.06.1986 +7 l in 06.03.1990	Aerobic	25.10.1984	2013
#26	$Uranium_{nat.}$ (951 g)	0.34	373	nat. $MgCl_2$-Solution	135	Aerobic	26.09.1984	2013
#27	$Uranium_{nat.}$ (926 g)	0.43	314	nat. NaCl-Solution	128 *33 l in 31.01.1989 +16 l in 03.04.1990	Aerobic	25.09.1984	2013
#28	^{137}Cs (370 MBq reference date 01.02.1984)	0.43	322	nat. $MgCl_2$-Solution	122 +30.9 l in 23.10.1984		26.09.1984	2006
#29	^{137}Cs (370 MBq reference date 01.02.1984)	0.43	322	nat. $MgCl_2$-Solution	125.3 l +3.2 l in 06.03.1990		26.09.1984	2013
#30	^{137}Cs (370 MBq reference date 01.02.1984)	0.43	322	nat. NaCl-Solution	124 l +7.2 l in 6.3.1990		26.09.1984	2006
#31	$Uranium_{nat.}$ (1017 g)	0.50	336	nat. NaCl-Solution	135 l	Argon	09.11.1988	2006
#32	$Uranium_{nat.}$ (1017 g)	0.50	336	nat. NaCl-Solution	135 l	Argon	09.11.1988	2013
#33	$Uranium_{nat.}$ (1017 g)	0.50	336	nat. $MgCl_2$-Solution	135 l +20 l in 17.07.1997	Argon	25.09.1989	2006

	Tracer	W/C ratio (l/kg)	Mass (kg)	Leachate	Volume of leachate (l)	Atmos-phere	Start of test	End of test
#34	$Uranium_{nat.}$ (1017 g)	0.50	336	nat. $MgCl_2$-Solution	135 l +20 l in 17.07.1997	Argon	25.09.1989	2013
#35	^{237}Np (3.70 MBq)	0.50	339	nat. NaCl-Solution	135 l	Argon	07.10.1987	2013
#36	^{237}Np (3.70 MBq)	0.50	335	nat. $MgCl_2$-Solution	135 l +25 l in 17.07.1997	Argon	25.09.1989	2013

The required $MgCl_2$-rich solution (Q-brine) was recovered directly from a brine pool of the Asse II salt mine [22]. Until 1984 the NaCl-rich solution was obtained by dissolution of natural rock salt. Since 1984 both solutions were recovered from the flooded Beienrode salt mine. The NaCl-rich solution was taken from a depth of 580 m, the Q-brine from 750 m depth. For the experiments with uranium and neptunium blocks #31 to #36 it was aimed to establish anaerobic conditions. Therefore, an argon gas atmosphere was applied with an over pressure of 20 mbar. The Ar atmosphere should prevent uptake of CO_2 from the air and the formation and precipitation of carbonate phases. The initial composition of the used solutions is shown in the following table (Tab. III).

Tab. III Initial composition of the solutions used for the full-scale experiments at the Asse II salt mine [42, 44].

Experiment	Tap water-systems #1, #6, #13, #14, #15		NaCl systems #2, #5, #10, #11, #12, #23		$MgCl_2$ systems #4, #7, #8, #9, #20, #21, #22, #24		NaCl from Beienrode #27, #31, #32, #35	$MgCl_2$ from Beienrode #25, #26, #28, #29, #33, #34, #36
	mol/L	±	mol/L	±	mol/L	±	mol/L	mol/L
Na^+	0.013	0.002	6.979	0.108	0.369	0.179	5.431	0.341
K^+	0.013	0.001	0.025	0.002	0.250	0.078	0.003	0.637
Ca^{2+}	0.002	0.001	0.012	0.002	0.003	0.005	0.022	0.000
Mg^{2+}	0.001	0.000	0.004	0.004	4.148	0.032	0.036	3.661
Cl^-	0.025	0.001	7.039	0.038	8.180	0.165	5.431	7.781
SO_4^{2-}	0.001	0.001	0.027	0.001	0.247	0.140	0.004	0.260
ρ /g ml^{-1}	1.004	0.002	1.196	0.005	1.303	0.008	1.210	1.320
pH	12.1	0.2	10.6	1.3	6.3	1.01	7.0	6.0

Errors of the ion concentrations are not available for the solutions from Beienrode.

2.4 Solution sampling and analytical procedures

Before 1998 the solutions were sampled and analyzed by GSF-IfT. Using the ducts and the installed tubing it was possible to obtain liquids from three vertical levels of the solution. In some of the sampling campaigns samples were also collected after mixing of the solution in contact with a cement block. Most of the analyses were directed to released tracers Cs, Np and U, i.e. their concentrations or activities in solution. A more complete analysis of the solution composition (including pH) was only occasionally determined. The results were indicated in several reports [18-22, 44, 45].

Since KIT-INE was in charge of the scientific responsibility for the full-scale experiments, the solutions were sampled in 1998, 2001, 2003, 2006, 2011 und 2013. Mixing of the solutions was no longer performed. Sampling of liquids from the annulus space around the cement blocks was performed using the tubings at levels of 30, 60 and 90 cm depth (referring to the top of the vessels). From each level, three syringes of 50 ml were filled, two of them filtered (450 nm). One of the filtered samples was acidified with about 0.05 ml nitric acid (supra, 65%). The third sample remained unfiltered. When possible, solids floating on the solution surface were sampled, or if easily removable, small samples from the surfaces of the blocks were collected. The density of the liquids was determined using a pycnometer.

At KIT-INE the dissolved cations in the sampled solutions were measured by ICP-AES, ICP-OES or ICP-MS. Due to the high ionic strength the leachates had to be diluted before analysis. Dilution factors for minor and major components were 2000 and 20000, respectively. For trace elements a dilution factor of 200 was applied. Dissolved anions were determined by ion chromatography, Cl^- by titration. Uranium was measured by ICP-MS, Np by high resolution ICP-MS and ^{137}Cs by radioactivity measurement with LSC. The main contribution to the errors of measurements resulted from the required dilutions and summed up to about ±10 %.

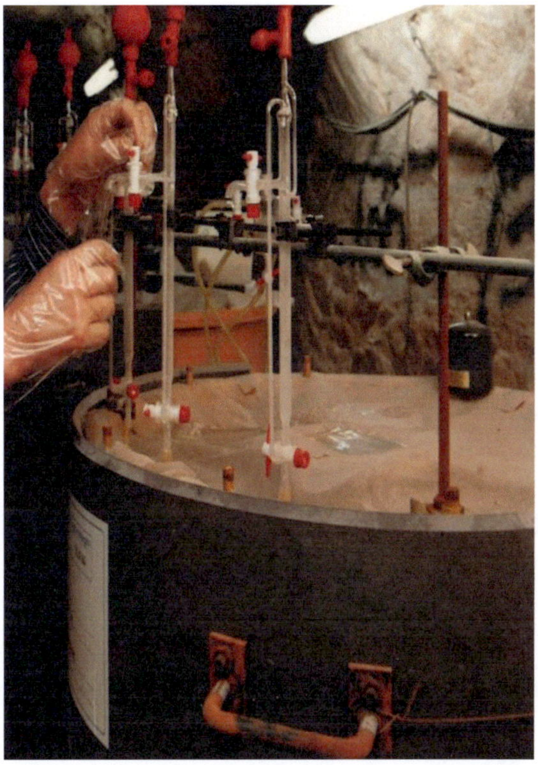

Fig. 5 Set-up for sampling of the liquids from the leaching vessels. Each of the three tubings samples from another vertical level in the solution. The canister in the image is surrounded by an additional lead shielding.

2.5 pH correction with respect to ionic strength

Measurement of pH values were done with a single-rod glass electrode (Type Ross, Model 920A, Orion). From pH measurements of diluted solutions ($I \leq 0.1$ mol/(kg H_2O)), the absolute activity coefficient of the dissolved hydrogen ion $\log \gamma_{H+}$ can be computed on the basis of the extended Debye-Hückel equations (± 0.02 log units). The molal H^+ concentration, m_{H+}, is defined by the following equation (a_{H+}: chemical activity of dissolved protons):

$$pH = -\log a_{H+} = -\log m_{H+} - \log \gamma_{H+} \tag{1}$$

For salt solutions of high ionic strength I ($I > 0.1$ mol/(kg H_2O)), the measured pH (pH_{exp}) is an operational value without physical meaning [46, 47]. The measured pH is related to the molal H^+ concentration according the following equation:

$$-\log m_{H+} = pH_{exp} + A \tag{2}$$

The parameter A is independent on the geometry of the Ross electrode. A is composed by contributions of $\log \gamma_H+$, the absolute (real) activity coefficient of the H^+ ions, and of $\Delta E_j \cdot F / (RT \cdot \ln 10)$. The difference between the different diffusion potentials is $\Delta E_j = (E(S) - E(X))$, resulting from the junction potential of standards and the solution, respectively. E(S) is determined by measurement of pH of standard buffers at low ionic strengths and E(X) by measuring in concentrated salt solutions. For a defined electrode system, bridging electrolyte and temperature, the parameter A depends only on the composition and concentration of the salt solution. Single ion activity coefficients $\log \gamma H^+$ can be calculated only at low ionic strength and not in salt solutions having $I > 0.1$ mol/(kg H_2O). Diffusion potentials can be calculated with high uncertainty. Therefore, the parameter A needs to be determined experimentally.

With a similar approach a directly measured pH_{exp} can be related to computed pH_{comp} when using the splitting convention:

$$pH_{comp} = pH_{exp} + \Delta pH \tag{3}$$

For experimental determination of the parameters A or ΔpH as a function of the NaCl, $MgCl_2$ and $CaCl_2$ molality, the pH_{exp} was measured in the respective salt solution having a known H^+ concentration [48, 49]. Taking these experimental values for $I > 0.1$ mol (kg $H_2O)^{-1}$, empirical polynoms for calculating A and ΔpH in NaCl, $MgCl_2$ and $CaCl_2$ rich solutions were derived. These polynoms allow to calculate molal H^+ concentrations ($-\log m_{H+}$) from measured pH_{exp}.

2.6 Solid sampling and analytical procedures

Due to the license for the storage room of the experiments at the Asse II salt mine sampling the radioactively doped simulates was not possible until 2006. Only some small samples from suspended material in the solutions could be obtained.

In 2006 KIT-INE was commissioned by the Federal Office of Radiation Protection (BfS) to investigate the properties and phase compositions of the corroded simulates. The contract covered analyses of the solids and interpretation of the results by comparison to findings from solution analyses. First, the corroding solutions were removed from cement blocks #28, #30, #31 and #33 and the corroded simulate blocks were stabilized by colored cement paste for the purpose of transportation [50]. Solid cores from the 4 simulates were recovered by dry drilling at the Research Center Jülich [50]. The simulate blocks investigated later were similarly sampled, by dry drilling at the HDB (Fig. 6) [28]. All drillings were performed from the top of the blocks and reached to a depth of 0.4 m (HDB) or 0.5 m (FZJ). The diameters of the drilling cores were 18 mm (HDB) or 45 mm (FZJ). The abraded drilling dust was collected separately for 10 cm intervals according to the drilling progress. The number of drilling cores and their positions, referenced to the rim of the cement blocks, are presented in Tab IV.

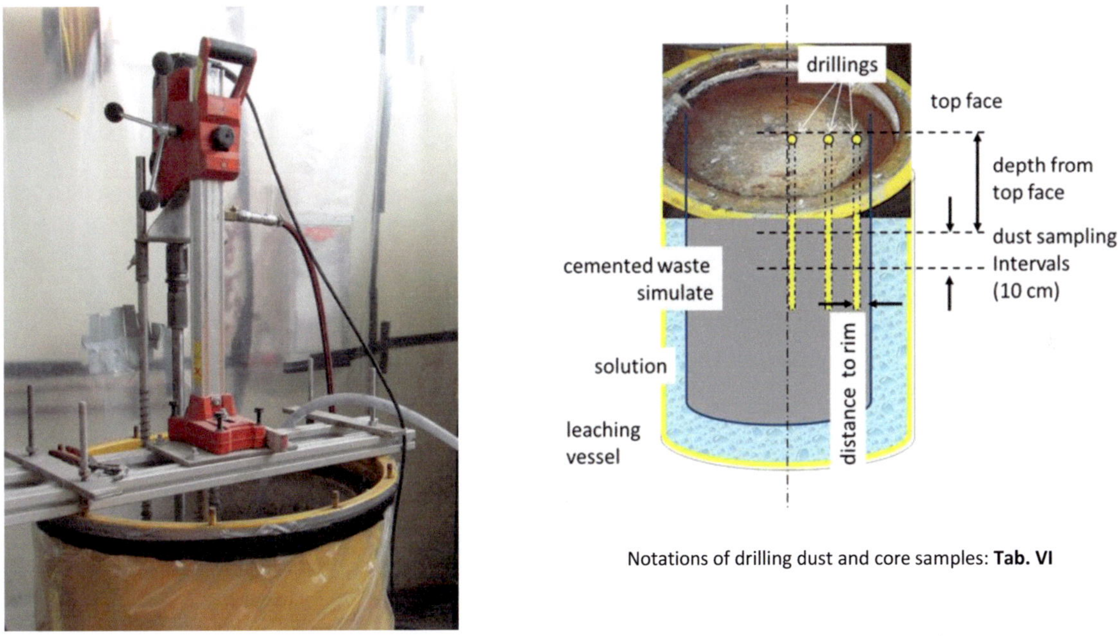

Notations of drilling dust and core samples: **Tab. VI**

Fig. 6 Set-up for dry drilling of radioactively doped cemented waste simulates in a hot cell at HDB and notations. The tube at the right side of the image was used for vacuum of the drilling dust.

In the subsequent contract with Helmholtz Zentrum München, Deutsches Forschungszentrum für Gesundheit und Umwelt GmbH, the formal owner of the experiments, the highest priority was attributed to investigations of neptunium retention. Until 2011, only the Np concentrations in the solutions have been analyzed. Furthermore, it was intended to test previous findings on U retention [30, 35] by investigating the U doped simulates of different W/C ratios. However; due to the experimental complexity of the investigations with the radioactive materials, only a limited number of solid samples could be analyzed. Remaining samples are stored at KIT-INE allowing further investigations. Beside the radionuclide containing drilling cores and drilling dust samples, KIT-INE has two drilling cores of one inactive Cs doped simulate (#14) which was leached in tap water.

Tab. IV Drilling cores and drilling dust material recovered from the simulate blocks.

Experiment	Number of drillings	Distance to the rim (mm)
#14	2	peripheral and central
#21	5	40, 80, 120, 150, 200
#22	2	40, 120
#23	3	40, 80,120
#24	2	80, 150
#25	5	40, 70, 100, 150, 200
#26	2	40, 120
#27	3	40, 80, 150
#28	7	75, 100, 135, 170, 195, 220, 255
#29	1	200
#30	3	60, 110, 120
#31	4	40, 95, 150, 210
#32	1	60
#33	3	80, 150, 210
#34	5	40, 60, 100, 120, 150
#35	5	40, 60, 100, 150, 200
#36	5	40, 60, 100, 150, 200

The drilling cores and/or drilling dust samples of full-scale cement blocks printed boldly in Tab. IV were investigated in details.

A flow sheet of the activities, investigations and applied methods in the scope of the solid sample characterization is shown in Fig. 7.

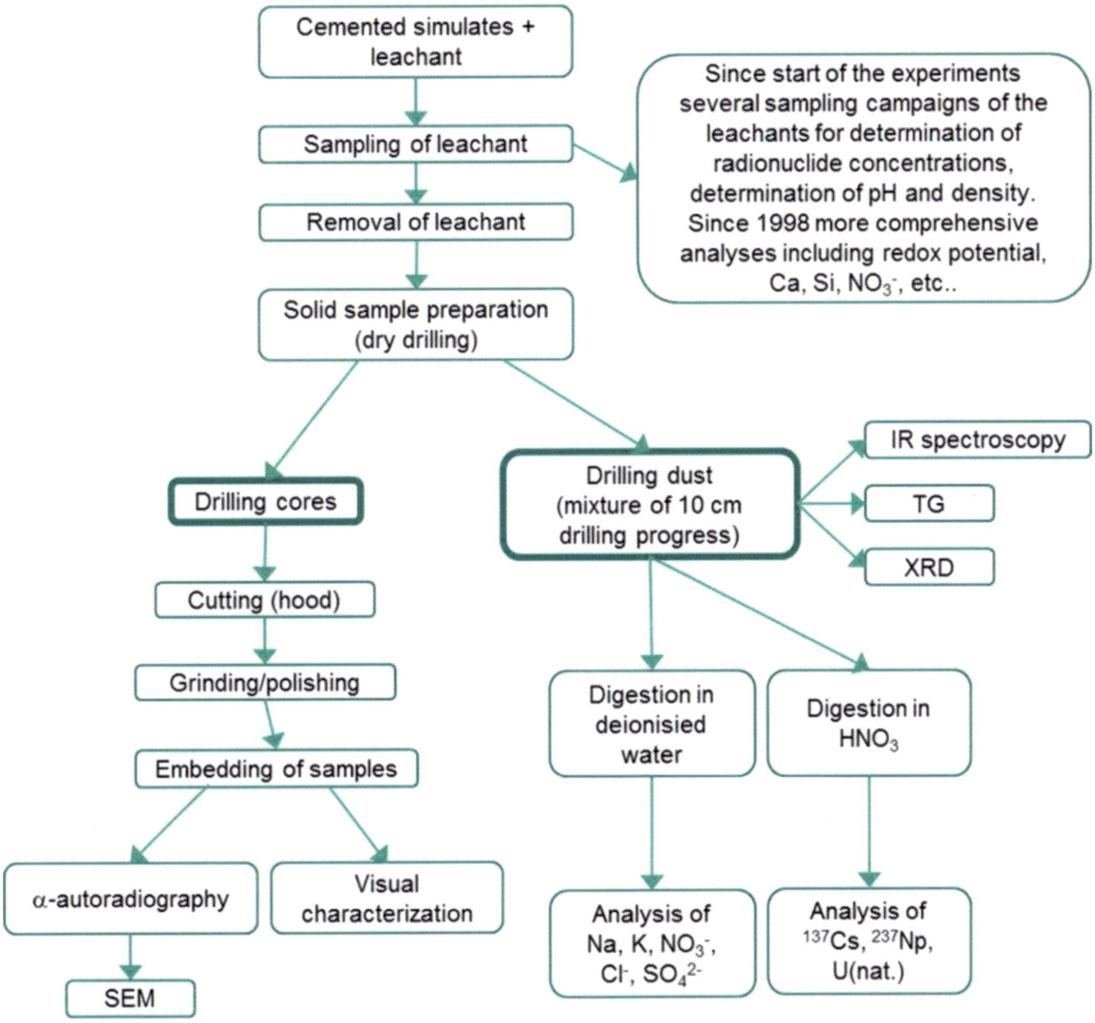

Fig. 7 Flow sheet of the activities, investigations and applied methods in the scope of the solid sample characterization and solution analysis.

2.6.1 Chemical analyses of cations and anions in the solids

The solids (core and drilling dust samples) were not completely digested, but extracts were prepared and analyzed.

Acid extracts: For quantification of the radionuclides, 0.5 g air-dried drilling dust was reacted with 25 ml of 4 % HNO_3 for 6 days and shaken regularly. These extracts were used for quantification of the neptunium (HR-ICP-MS), the uranium and the cesium content (ICP-MS) in the corroded solids.

Aqueous extracts: 0.5 g air-dried drilling dust was reacted with 25 ml deionized water for 14 days and shaken regularly. The aqueous extracts were used for determination of the more soluble compounds sodium and potassium (ICP-OES) as well as nitrate and chloride (IC) in the corroded solids.

Aqueous extracts allow the quantification of nitrate in the solids in the case that nitrate exists in an easily soluble salt. Measured chloride content in the cement product depends on the mineral phases of the solid. Only the part of chloride in the solid soluble in water can be determined. Sulfates will be rarely dissolved.

The pH of the extracts was not determined, but it can be assumed that where portlandite was present it buffered the pH at 12.5 in the aqueous extracts. In the case of portlandite free samples, or if portlandite was consumed, the pH was controlled by C-S-H phases in the approximate range 10.5 to 12.5.

2.6.1.1 Trace analytics (HR-ICP-MS and ICP-MS)

The neptunium concentrations of the acid extracts were determined by means of High Resolution (magnetic sector) ICP-MS. The accuracy of the calibration was tested by measurement of reference materials (TM26.4 for ^{238}U, certified solution from Eckert & Ziegler Nuclitec GmbH for ^{237}Np). The standard deviation of the ^{237}Np measurements was below 3 % for all analyzed samples.

The uranium concentrations in the acid extracts were measured quantitatively by a quadrupole ICP-MS Elan6100, adapted to a glove box. The blank value and the calibration standards were prepared with 2 % HNO_3 (ultrapure). The blank samples, calibration standards, reference standards and the samples were spiked by 10 μg/l rhodium as an internal standard. The samples were measured in the sequence given above. The standard deviation of single uranium measurements was below 4 % for all samples.

From 1984 to 2013, the uranium concentrations in the leachates were determined by various methods. After fast initial dissolution of a small fraction of the uranium, the measured aqueous concentrations showed no decreasing or increasing trend. To evaluate the scatter of all measurements over the whole experimental period, all documented data were used. In the case of solution from experiment #25, the scatter amounted to ± 32 %. Since 2003 the uranium analyses were performed by ICP-MS. These analyses showed a scatter of ± 16 % between different sampling campaigns.

2.6.1.2 Analysis of cations (ICP-OES)

Main cations were determined by ICP-OES. This device is a simultaneous spectrometer using a spectral range from 180 to 800 nm for element identification. The instrument was used for sodium and potassium quantification in diluted aliquots of the aqueous extracts.

2.6.1.3 Analyses of anions (IC)

The anion concentrations in the aqueous extracts were determined by ion chromatography with an ICS-1100 instrument and an anion exchanger column AS9-HC (Thermo Fischer). The column was selected because of the separation of nitrate and bromide. Bromide might be present in natural salt solutions.

Filtration (450 nm) did not show an effect on measured nitrate concentrations. Based on this experience all anion measurements were performed with unfiltered leachate solutions.

The error of the concentrations determined in the aqueous extracts was below ± 15 % for chloride and below ± 10 % for nitrate (aqueous extract from block #25 and #35).

2.6.2 Solid phase analytics

The characterization of solid material was mainly done with drilling dust samples, collected at different depths. However, these analyses were complemented by microscopic investigations of undisturbed core samples analyses which allow a comprehensive interpretation of the radionuclide behavior.

The distributions of uranium and neptunium were analyzed at discrete depths of the drilling cores. For these investigations the samples were prepared specifically and first analyzed by means of non-destructive α autoradiography. Afterwards, macroscopic and microscopic investigations of these samples were performed by digital microscopy and scanning electron microscopy (SEM). The different techniques revealed information on the porosity, mineral paragenesis and mineral composition, and provided hints on the retention or mobilization mechanisms of radionuclides. Important analytical techniques are briefly introduced in the following chapters.

2.6.2.1 Element analyses by X-ray fluorescence spectroscopy (XRF)

In addition to the analyses of the extracts, element analyses of the drilling dust samples have been performed using XRF. For the calibration of degraded cement materials neither certified reference materials nor comparable natural or technical samples exist. For this reason, own calibration standards consisting of mixtures of element oxides and chlorides were prepared.

2.6.2.2 X-ray powder diffractometry (XRD)

To identify mineral phases, the pdf-2 data base (release 2000) from the International Centre for Diffraction Data together with the Bruker Eva-Software Version 3.0 were used. XRD is an important technique to characterize the corroded cement paste, but is complicated by the multitude of different phases present in the samples and was therefore complemented by other solid-phase characterization techniques. Identification by XRD is especially hindered by several, possibly present mixed phases or solid solutions with varying lattice spacings. Furthermore, some hydrates change lattice spacings not only due to exchange of anions in interlayers, but also according to the drying procedure of the sample due to removal of interlayer water. Moreover, hydrated phases and persistent clinker phases show extensive overlap of reflexes.

2.6.2.3 Thermogravimetry / Differential Thermal Analysis (TG/DTA)

Measurement of exothermal and endothermal reactions as well as mass loss during heating of small powder samples allows identification of some minerals and under optimal conditions also their quantification. The results of thermogravimetric measurements were partly affected by the presence of salt in some solid samples. The TG/DTA, in some cases combined with gas analyses, was mostly applied for quantification of free water in the corroded cement phases, and verification of the presence of portlandite, nitrate and carbonate.

2.6.2.4 Scanning electron microscopy (SEM)

For evaluation of the topography of a sample, mainly secondary electrons originating from the sample are used. These secondary electrons are formed in a thin layer at the surface (<1 µm) and therefore provide high resolution topographic information of the sample surface. Backscattered electrons (BSE) provide information from depth in the sample (some µm, depending on the electron beam density and the acceleration voltage). The yield of these electrons depends, among other factors, on the atomic number of the elements. Heavy elements with more electrons scatter more electrons and therefore give higher intensity in the detector. Color contrasts of flat samples in BSE-mappings indicate the distribution of heavy elements in the sample surface. For quantification (without standards) the emitted, element specific X-Rays are recorded by an EDX detector from preferably plane samples. In this way spot analyses, line scans or two dimensional 'maps' of the element distribution at the surface of samples can be recorded.

The electron microscope was equipped with a 'low-vacuum mode'. This method has benefits for analyzing C-S-H phases, which partly release structural water and change their structure under high vacuum conditions.

2.6.2.5 Infrared spectroscopy (IR)

Infrared (IR) spectroscopy was performed using an ATR (Attenuated Total Reflection) Spectrometer equipped with a DTGS detector. For IR spectra 64 scans were recorded ranging from 4000 to 400 cm^{-1} with resolution of 4 cm^{-1}. For the ATR measurements a MIRacle ATR Diamant Cell was applied. The powdered sample was pressed in the cell using a sapphire punch.

This device is available at KIT-INE for non-radioactive samples, only. For elucidation of the nitrate binding in the cement structure inactive samples from cemented simulate #14 were investigated.

2.6.2.6 Radiological measurements

^{137}Cs in the solid (drilling dust) was determined by γ spectrometry. About 1 g drilling dust was filled in a plastic vial and adjusted to a Canberra HPG3 detector. The measurements were not quantitatively calibrated, but equally measured samples are comparable with each other.

Autoradiography measurements were performed using a Cyclone Plus Storage Phosphor System. For measurements a 'multisensitive Film' was covered by a thin, transparent film, in order to avoid any contamination. The solid samples were placed on the film together with a radioactively doped reference. During exposure of the film to samples for hours or days, it was protected from light to reduce the background. Afterwards, the film was scanned electronically to measure the two dimensional intensity distribution in the samples, resulting mainly from α-particles

The full-scale simulates were doped by $2.44 \cdot 10^7$ Bq natural uranium (experiment #25) and $3.7 \cdot 10^6$ Bq ^{237}Np (experiments #35 and #36). α-particles have a range of about 10 µm in cement paste. This means that only those α-particles coming from this 10 µm layer can reach the film. A core of 2 cm diameter accounts for an activity of 0.47 Bq (uranium) and 0.071 Bq (^{237}Np) in a 10 µm layer. Within an exposition of 18 hours about 10^4 α decays (uranium) and 10^3 α decays of ^{237}Np might be registered. However, the transparent film provides for an additional reduction of the radiation and only a fraction of the radiation is emitted in the direction of the film.

2.6.2.7 Time resolved laser fluorescence spectroscopy (TRLFS)

TRLFS was applied to drilling dust samples. TRLFS measurements were performed using a pulsed Nd:YAG pumped dye laser system at a repetition rate of 10 Hz and a maximum laser energy of 3.5 mJ. The description of the laser systems is published in detail [51]. The powders were illuminated by a pulsed laser and fluorescence light was recorded by fiber optics perpendicular to the incident laser beam. The die excitation wave length λ_{ex} was 400 nm (laser dye: Exalite 398), the delay time after excitation 100 ns or 1 µs the gate width for the fluorescence measurements 100 ns for short delay time and 1 ms for the extended delay time.

2.6.2.8 X-ray absorption spectroscopy XANES/EXAFS

To determine the chemical environment of the uranium species in the solids, core and dust samples were analyzed by X-ray absorption fine structure spectroscopy at the INE-Beamline for actinide research at the ANKA [52]. The INE-Beamline is designed particularly for actinide research by means of X-ray absorption spectroscopy. The beamline is operated by INE and is equipped with a radioactive instrument hut. The beamline is classified as "Pooled Facility" within the scope of the EU-FP-7-Project TALISMAN. Investigations with non-fissile radionuclides are licensed up to 10^6 times of the exemption limit and fissile isotopes (^{239}Pu, ^{235}U) up to 200 mg. XANES spectra were recorded including reference compounds, such as Ca-uranate, Na-uranate, meta-schoepite, soddyite and uranophane.

2.6.2.9 Raman spectroscopy

By means of Raman spectroscopy molecular vibrations were recorded which provide evidence of crystallinity, composition and impurities or dopants. Raman spectroscopy makes use of the so called Raman effect which modifies the wavelengths and intensities of inelastically scattered light. The low-energetic light at wavelengths of 5 - 300 cm^{-1} is generated by molecular or crystal lattice vibrations. The scattered light has lower energy than the incident beam, i.e. it has a lower frequency. The observed spectral lines are termed Stokes lines.

3 Results and Discussion

3.1 Visual findings

Before start of the leaching experiments, photos of the full-scale blocks were taken (Fig. 4). Some blocks showed inhomogeneous appearance at the cylindrical surface especially in the lower part of the blocks. It is reasonable to assume that in these parts the cement suspension were not sufficiently mixed during the preparation of the simulates.

Some blocks were visually examined during the experiments: In 1983 inactively doped blocks were removed from the leachates for a short time and inspected. In most cases only some precipitates on the surfaces could be seen whereas indicative features of intensive degradation, such as fractures, were not present. This was also the case with block #23 (W/C ratio 0.32 l/kg) after 3 years in $MgCl_2$-rich solution. However, in the case of simulate #29 (W/C ratio 0.43 l/kg) already after two years in $MgCl_2$-rich solution, development of fractures was observed. In April 1992, cement blocks both in NaCl- and $MgCl_2$-rich solutions (W/C ratio 0.24 l/kg) visually still did not show obvious corrosion effects. On the top surface of block #27 halite crystals precipitated. In October 1993 two blocks exposed since 1979 to NaCl- and $MgCl_2$- rich solution (W/C ratio 0.24 l/kg) were inspected. Also in these cases only some shrinkage cracks were observed at the surface, but no hints on cement degradation [19]. After 10 years leaching in $MgCl_2$ solution cement blocks #33 and #34 (W/C-ratio 0.5 l/kg) showed a strongly structured surface and thick coatings attached to the surfaces. The coating thickness increased with time. After about 18 years block #33 was investigated in detail [34]. The composition of the coatings was analyzed and is documented in the report FZKA 7059 [26].

After removal of the solutions in 2006 and 2013 significant differences were found regarding the visual characteristics of the cement blocks:

- The blocks #13, #14 and #15, stored in tap water, appeared to be unchanged after 34 years of exposure. The W/C ratio of these simulated amounted to 0.24 l/kg.
- The blocks #20 to #26, stored in both - $MgCl_2$- and NaCl-rich brines, showed no visually detectable hints of degradations. On the top surfaces precipitates of different thicknesses were present. The W/C ratios lay between 0.30 l/kg and 0.34 l/kg.
- The cement blocks produced at higher W/C ratios (#27 at 0.43 l/kg, #30, #32 and #35 at 0.50 l/kg) corroded in NaCl-rich solution did not show significant changes. Precipitated coatings of different thicknesses could be seen on the surfaces, but no deep fractures into the cement matrix were visible (Fig. 8).
- The blocks produced at higher W/C ratios (#28 and #29 with 0.43 l/kg, #33, #34 and #36 with 0.50 l/kg) which were stored in $MgCl_2$-rich brine appeared to be largely degraded. Deeply penetrating fracture structures were present and the blocks appeared to be mechanically instable (Fig. 8).

Hence, the obviously different durability depended on the different W/C ratios and the initial composition of the solutions.

Full-scale block #36 after 23.3 years exposure in $MgCl_2$ solution

Full-scale block #33 after 17.1 years exposure in $MgCl_2$ solution

Full-scale block #35 after 25.3 years exposure in NaCl solution

Full-scale block #32 after 26.7 years exposure in NaCl solution

Fig. 8 Visual appearance of full-scale cement blocks (W/C ratio 0.5 l/kg) after long-term exposure to $MgCl_2$ (#33 and #36) or NaCl solution (#32 and #35). Red material is died, hardened cement paste, used for transport stabilization.

3.2 Leachates

In the case of a contact between cemented waste blocks and water or salt brines, concentration gradients exist between the bulk solution and the solution in the pore space. These concentration gradients force diffusively controlled exchange processes. Both, the basic literature [53-55] as well as the experimental investigations at INE described the following major corrosion mechanisms:

- Leaching: This term describes the mobilization of radionuclides such as cesium, and other easily soluble components from the hydrated cement product. In parallel, mobile species (e.g. Cl^-) from the solution diffuse into the cement. As a consequence of the leaching process, solid portlandite ($Ca(OH)_2$) can be dissolved leading to an increase of the porosity of the cement product and to an increase of the pH in the bulk solution. Similarly, the solidified $NaNO_3$ load easily dissolves and does not reprecipitate, i.e. porosity is increased.
- Calcium-Magnesium exchange reactions: This corrosion process is characterized by the following, simplified reaction (1).

$$MgCl_2(aq) + Ca(OH)_2(cr) \rightarrow CaCl_2(aq) + Mg(OH)_2(cr) \quad (1)$$

Brucite ($Mg(OH)_2$) precipitates in a flaky form. Formation of magnesium silicates may also occur. In the concrete industry these precipitates are considered as protective layer against further Mg^{2+} attack [56]. The process is considered in particular effective for relatively dense cement materials.

- Crystallization pressure: A mechanically detrimental effect may be exerted by the formation of secondary solid phases having a higher specific volume and possessing an elevated crystallization pressure. This type of corrosion may occur, e.g., where sulfate attack occurs.

In the case of a big excess of reacting ions in the solution or in the case of solution exchange, the pH may be reduced strongly (pH < 10), disturbing the stability of C-S-H phases. Further decrease of pH would give rise to newly formed secondary phases with completely different chemical, mineralogical and mechanical properties.

Several parameters which can be detected by analyses of the leachate are suitable for assessment of the corrosion progress. In the presented investigation, following quantities were mainly regarded:

- pH
- Nitrate concentration
- Ca, Mg and sulfate concentrations

3.2.1 Evolution of the leachates

The following sub-chapters show the temporal evolution of relevant quantities of the leachates compositions, summarized for the different systems under investigations. The graphs show measured values which are not distinguished according to all aspects of the experiments, but only according to the actual W/C ratios and the types of corroding solutions.

The data of the solution analyses can be traced to the previous reports [18-20, 25, 26, 28, 42, 45, 57, 58]. The detection limits of main and trace components of the leachates are given in Tab. V.

Tab. V Dilution factors and detection limits in the scope of the solution analyses at KIT-INE

	Dilution factor	Detection limit mmol/l
Ca	2000 / 20000	5.0 / 0.5
Na	2000	8.7
Si	200	1.0
Cr	200	0.4
Fe	200	0.3
Al	200	3.7
K	2000	5.1
Mg	2000 / 20000	8.2 / 0.8
SO_4	200	1.0
Cl	No because of titration	2.8
NO_3	200	1.8

Due to the high dilution factors which were applied stepwise, the error in measurement is estimated to ±10%.

3.2.1.1 pH

One of the most important, characteristic values of leachates is the pH. The evolution of the pH as function of time proceeded differently in tap water and NaCl-rich solution systems compared to $MgCl_2$-rich solution systems (Fig. 9): In the NaCl systems an increase of pH takes place within weeks starting from a neutral solution to values between 11 and 12.5. This increase is independent on the W/C ratio and on the ratio of cement mass to the volume of the leachate, which varied between 0.7 kg/l and 3.0 kg/l in different experiments.

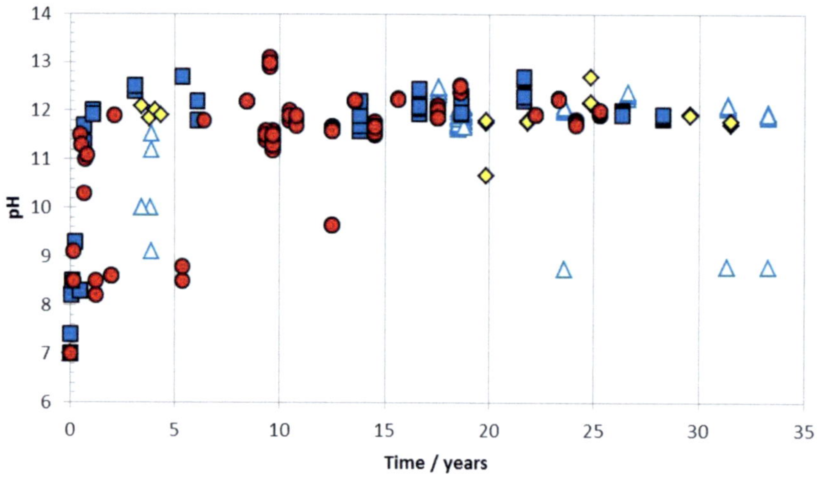

a.) NaCl-rich solution systems

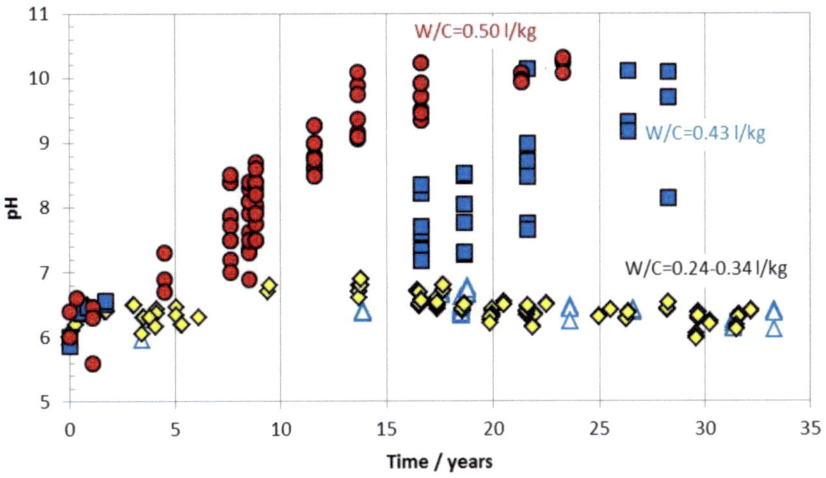

b.) $MgCl_2$-rich solution systems

Fig. 9 Temporal evolution of measured pH values in the (a) NaCl-rich solution and (b) in the MgCl2-rich solution systems, sorted by W/C ratios of the cement blocks.
Error of pH measurements was in the range of ± 0.1 pH unit.
Triangles: W/C ratio 0.24 l/kg, diamonds: 0.3 ≤ W/C ratio ≤ 0.43 l/kg, squares: W/C ratio = 0.43 l/kg, circles W/C = 0.5 l/kg

In $MgCl_2$-rich solution systems with cement blocks having a low W/C ratio the solutions remained in the initial pH range. In the case of block #7 to #26 with a W/C ≤ 0.34 only limited amounts of magnesium from the leachates was consumed until termination of the experiments and the pH was stable at 6 ≤ pH ≤ 7. This observation can be explained by precipitation of Mg hydroxides at the surfaces of the blocks and the correlated inhibition of further corrosion [56]. In contrast to that, the pH of solutions from experiments #28, #29 #33, #34 and #36 with realistic W/C ratios evolved with time. In these cases, magnesium from the solutions was consumed by the Ca-Mg exchange reaction and pH increased to values above 10, due to dissolution of alkali hydroxides, portlandite and C-S-H.

3.2.1.2 Evolution of aqueous Ca and Mg concentrations

The temporal evolution of the calcium and magnesium concentrations in the NaCl- and $MgCl_2$-rich solution systems is shown in Fig. 10 and Fig. 11. The data are sorted according the W/C ratios of the full-scale cemented simulates. In NaCl brine the calcium concentration remains at the initial level. After 20 years, a slightly decreasing tendency exists. The Mg concentrations in the NaCl systems were initially in the range of 0.035 mol/l. First measurements after 6 to 11 years showed slightly decreasing values, however, a significant scatter of the data is present. The reasons for the scatter cannot be clarified anymore.

Corrosion and leaching of easily soluble components, can be tracked by the concentrations measured in solution. Moreover, in the case of $MgCl_2$-rich solutions the corrosion process can be quantified due to the occurring Ca-Mg exchange reactions (Fig. 10b and Fig. 11b). Cement blocks with W/C ratio 0.5 and 0.43 l/kg showed an increase of the Ca concentration and a decrease of the Mg concentration. In both cases, the average final Ca concentration varied between about 2 and 3 mol/l, the Mg concentrations decreased to values close to the detection limit. However, experiments with blocks of W/C ratio 0.5 and 0.43 l/kg differ from another in the the kinetics of the observed Ca-Mg exchange. The intersection of the increasing Ca and the decreasing Mg curves for blocks of W/C = 0.5 l/kg was found after about 7.5 years, whereas for blocks with W/C = 0.43 after 15.5 years. Blocks with even lower W/C ratios did not show an intersection within the experimental period up to 33 years.

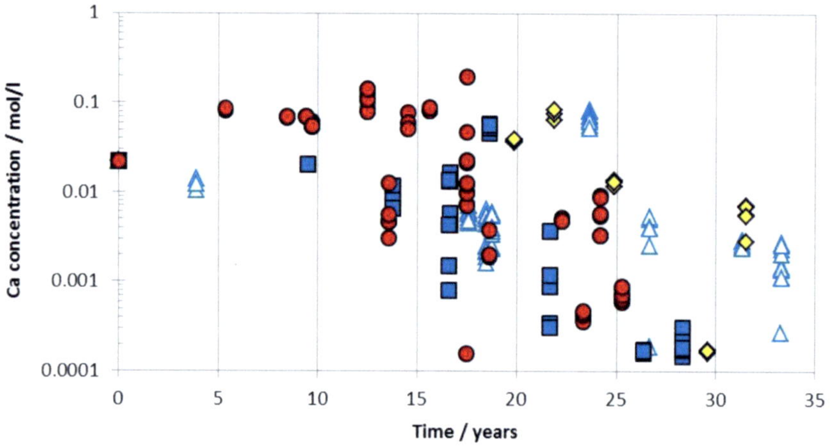

a) NaCl-rich solution systems

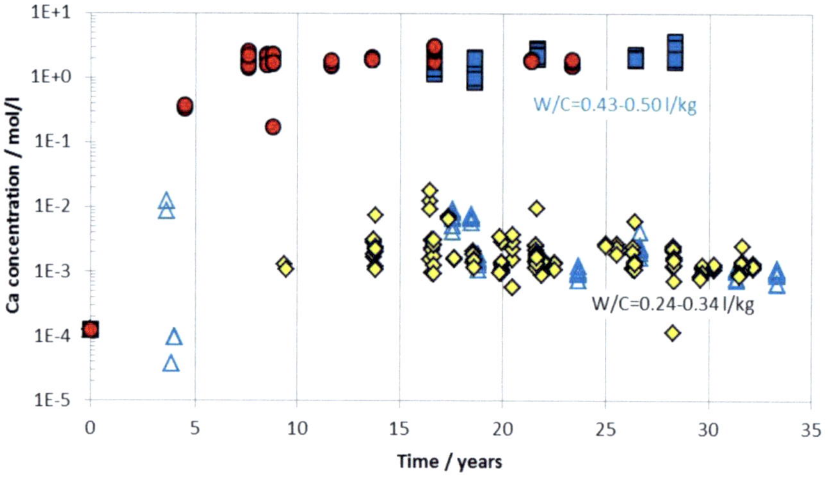

b) $MgCl_2$-rich solution systems

Fig. 10 Temporal evolution of measured calcium concentrations in the (a) NaCl-rich solution systems and (b) in the $MgCl_2$-rich solution solution systems, sorted by W/C ratios of the cement blocks.
Triangles: W/C ratio 0.24 l/kg, diamonds: 0.3 ≤ W/C ratio ≤ 0.43 l/kg, squares: W/C ratio = 0.43 l/kg, circles W/C = 0.5 l/kg

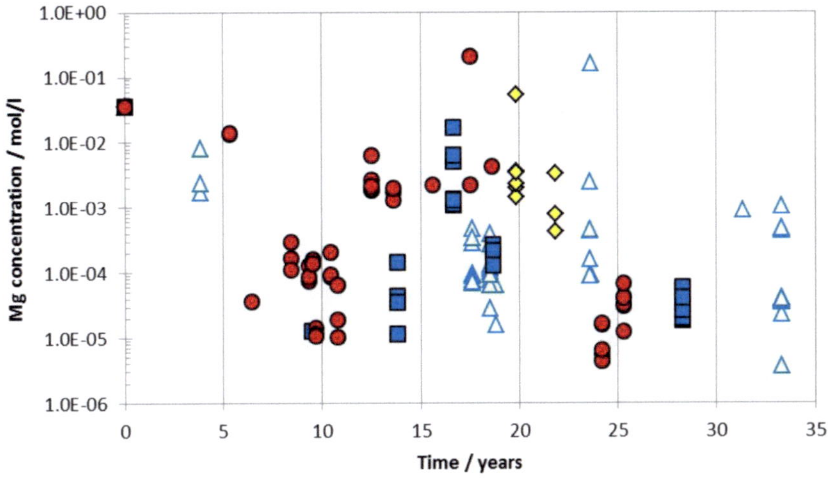

a) NaCl-rich solution systems

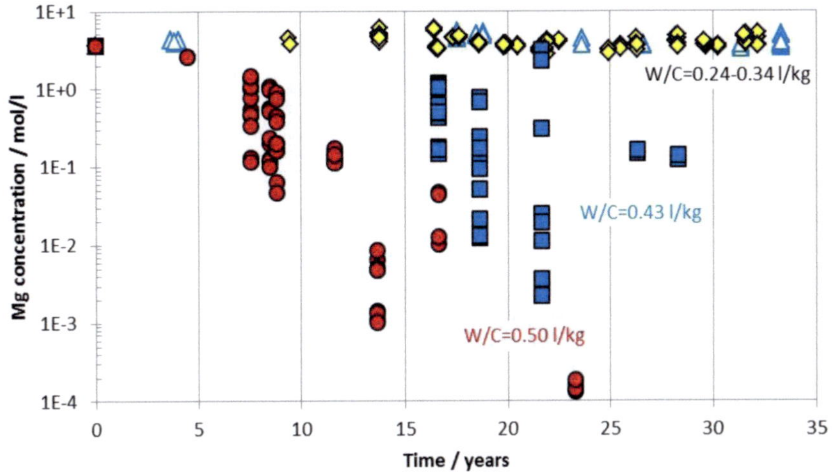

b) $MgCl_2$-rich solution systems

Fig. 11 Temporal evolution of measured magnesium concentrations in the (a) NaCl-rich solution systems and (b) in the $MgCl_2$-rich solution systems, sorted by W/C ratios of the cement blocks.
Triangles: W/C ratio 0.24 l/kg, diamonds: $0.3 \leq$ W/C ratio ≤ 0.43 l/kg, squares: W/C ratio = 0.43 l/kg, circles W/C = 0.5 l/kg

3.2.2 Evolution of the densities of the leachates

In the NaCl systems the leachate's density hardly showed changes over the experimental period. In contrast, in the $MgCl_2$-rich leachates a certain change of the leachate's density was observed, depending on the corrosion progress of the cement blocks. Fig. 12 shows the evolution of the densities of the initially $MgCl_2$-rich solutions as function of time for the different W/C ratios. The densities of the initially $MgCl_2$-rich solutions are given in Tab. III, however the experimental protocols showed partly different values. Fig. 12 shows the evolution of average solution densities as a function of the duration of the experiments. For experiment #24 a solution of higher initial density was applied (triangles) remaining at a higher level until the end of the experiment. For the other cement blocks with W/C = 0.32 kg/l, a certain decrease of the densities was observed in the beginning of the experiments. For cement blocks with W/C = 0.5 l/kg, the densities increased again after about 5 years, and were finally stabilized at a density level of about 0.02 kg/l below the initial density. The error bars shown in Fig. 12 represent the deviations of solution samples collected at different levels of the experimental vessels.

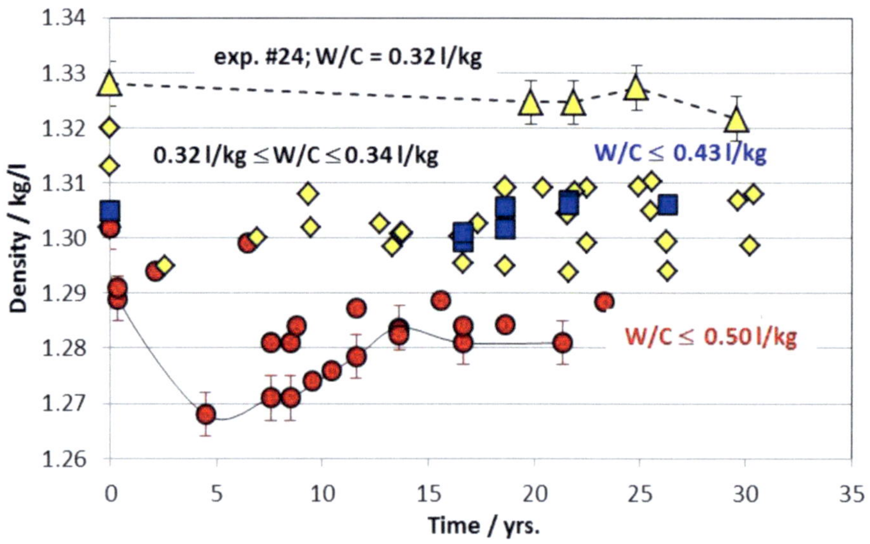

Fig. 12 Evolution of the density of initially $MgCl_2$-rich solutions in contact with simulate blocks of different W/C ratios.
Dashed line: Experiment #24 having a higher initial density
solid line: Experiment #34 (W/C = 0.5 l/kg)

3.2.3 Mobilization of waste components

The following graphs show the temporal evolution of waste component concentrations released in the NaCl-rich and $MgCl_2$-rich leachates. The waste components which are discussed here comprise cesium, neptunium, nitrate and uranium.

3.2.3.1 Cesium and nitrate

The share of inactive Cs which was mobilized from the added inventory was calculated from the average concentrations determined in the solutions. The average was calculated from the measured concentrations at the sampling positions 'top', 'middle', 'bottom' in the leachate vessels. In the case of the radioactively doped cement blocks, the average concentrations and the initial inventory were corrected with respect to the radioactive decay (half-life of ^{137}Cs is 30.17 years).

In the $MgCl_2$-rich solutions the low W/C ratio blocks (≤ 0.34 l/kg) showed a relatively low Cs mobilization during the whole observation period (Fig. 13). The fraction of mobilized cesium from all cement blocks at the end of the experiments is listed in Tab. VI.

Tab. VI Fraction of Cs mobilized into the leachate at termination of the experiments.
^{137}Cs data: corrected according to the radioactive decay.

Block ID	Dopant	Initial Inventory	Key date radioactivity measurement	Fraction of radionuclide found in the leachate %	Duration of the experiment years
#7	Cs(inactive).	223 g	-	10.9 ± 3.3	33.3
#8	Cs(inactive).	233 g	-	8.8 ± 2.3	33.3
#9	Cs(inactive).	236 g	-	6.8 ± 2.8	33.3
#10	Cs(inactive).	246 g	-	51.1 ± 2.9	33.3
#11	Cs(inactive).	253 g	-	32.3 ± 7.1	33.3
#12	Cs(inactive).	253 g	-	39.8 ± 4.1	33.3
#13	Cs(inactive).	231 g	-	42.5 ± 5.8	33.3
#14	Cs(inactive).	232 g	-	40.1 ± 5.3	33.3
#15	Cs(inactive).	237 g	-	34.1 ± 6.0	33.3
#20	Cs(inactive).	509 g	-	12.0 ± 2.2	32.2
#21	^{137}Cs	100 mg $CsNO_3$ / $3.42 \cdot 10^8$ Bq	27.10.1980	7.4 ± 0.4	31.6
#22	^{137}Cs	100 mg $CsNO_3$ / $3.42 \cdot 10^8$ Bq	27.10.1980	7.9 ± 0.4	31.6
#23	^{137}Cs	100 mg $CsNO_3$ / $3.44 \cdot 10^8$ Bq	27.10.1980	49.8 ± 1.2	31.5
#24	^{137}Cs	$6.85 \cdot 10^8$ Bq	27.10.1980	5.5 ± 0.7	31.5
#28	^{137}Cs	0.42 mg Cs / $3.70 \cdot 10^8$ Bq [1]	01.02.1984 (estimated)	85.3 ± 5.8	22.1
#29	^{137}Cs	0.42 mg Cs / $3.70 \cdot 10^8$ Bq	01.02.1984	73.5 ± 5.6	28.3
#30	^{137}Cs	0.42 mg Cs / $3.70 \cdot 10^8$ Bq	01.02.1984	34.8 ± 3.5	22.1

Cemented blocks of W/C ratio 0.43 l/kg in $MgCl_2$-rich solution showed a similar mobilization behavior as blocks of lower W/C until about 7 years of experiments. Then, the mobilization process changed drastically. In the following 10 years period substantial Cs mobilization occurred and leveled off at about 80 % of the inventory.

For cement blocks corroded in NaCl-rich solution, the Cs mobilization was apparantly independent on the W/C ratios (Fig. 13a). The diffusion controlled mobilization in the NaCl system continued over the whole experimental duration. In total, after more than 30 years, about 40 - 50% of the Cs inventory was released to the solution.

1 Different masses of the cesium inventory of block #28 - #30 have been recorded [13]. There is a suspicion that the data refer to different Cs compounds used for doping the cement blocks. The compounds have not been specified, therefore the percentage of leached Cs bases on the radioactivity.

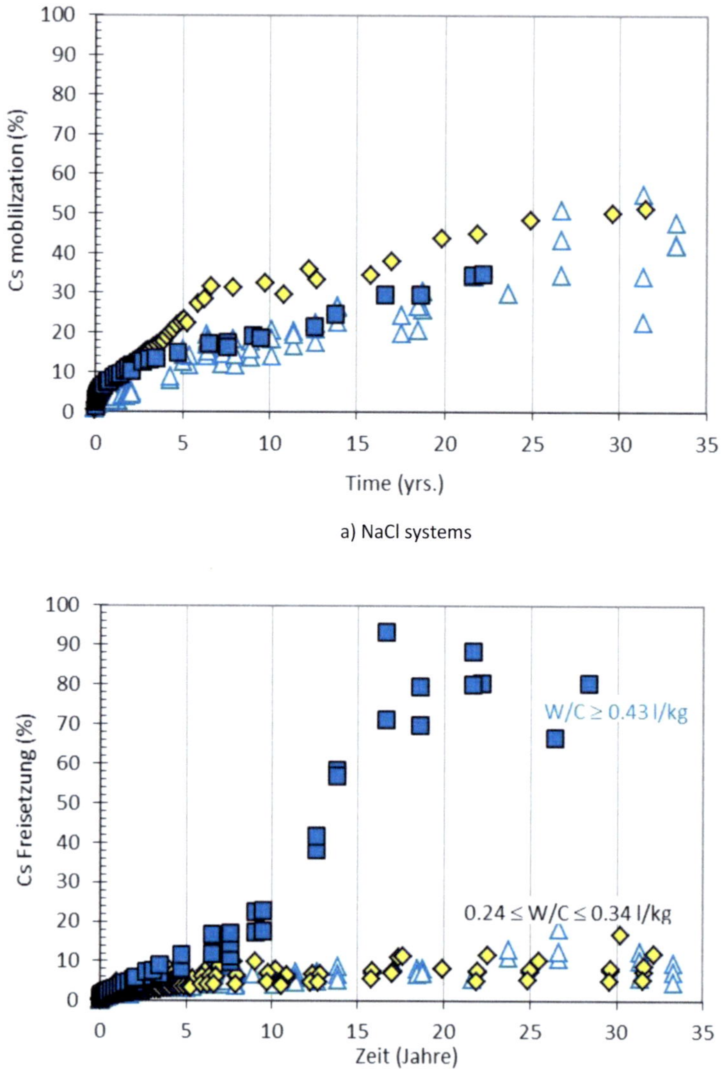

a) NaCl systems

b) $MgCl_2$ systems

Fig. 13 Cesium release as function of time (related to the added inventory of the dopant) in (a) NaCl- and (b) $MgCl_2$-rich solution systems, sorted by W/C ratios of the cement blocks.
Triangles: W/C ratio 0.24 l/kg, diamonds: 0.3 ≤ W/C ratio ≤ 0.43 l/kg, squares: W/C ratio ≥ 0.43 l/kg

The nitrate concentration was not measured in the leachates during the first years of the experiments. Therefore, only concentration data since 2003 can be reported (Fig. 14). The mobilization of nitrate is similar to the Cs release. In $MgCl_2$ systems, cement blocks having high W/C ratios showed significant release, whereas for low W/C ratio blocks very low amounts of nitrate were found in the leachates. In NaCl systems, the fraction of mobilized NO_3 was less in comparison to Cs, however, only few data support this finding.

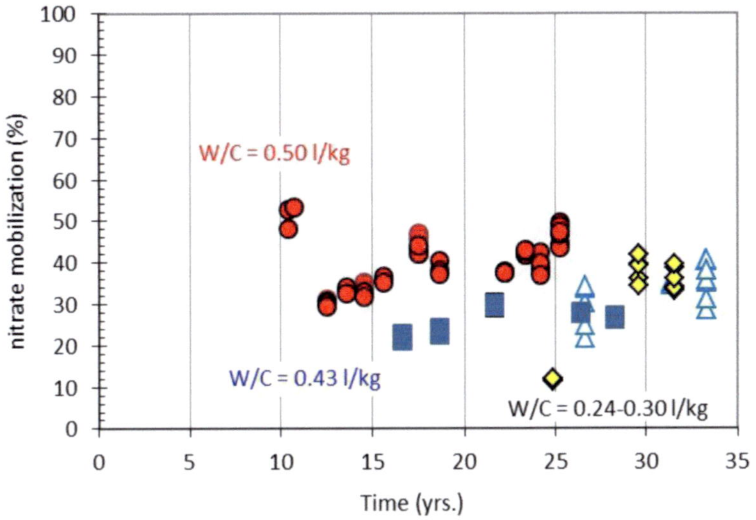

a) NaCl system

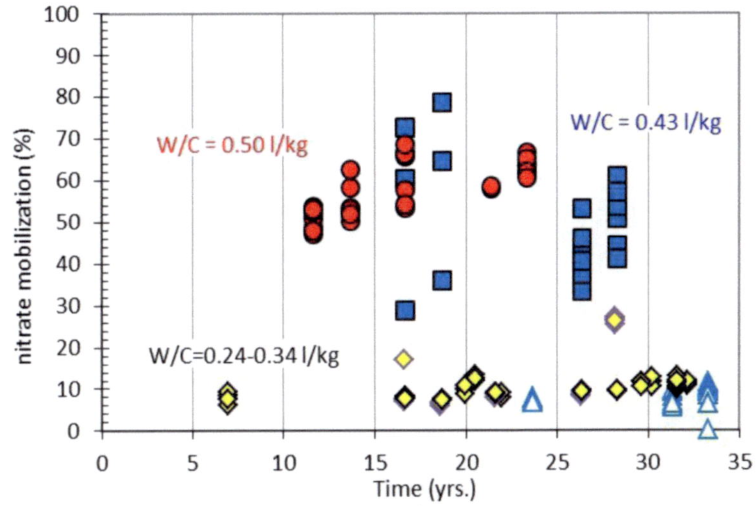

b) $MgCl_2$ systems

Fig. 14 Nitrate release as function of time (related to the initial mass/activity of the dopant) in (a) NaCl and (b) $MgCl_2$ solution systems, sorted by W/C ratios of the cement blocks.
Triangles: W/C ratio 0.24 l/kg, diamonds: $0.3 \leq$ W/C ratio ≤ 0.43 l/kg, squares: W/C ratio ≥ 0.43 l/kg

3.2.3.2 Uranium

The mobilization of uranium related to the inventory was found to be very low for all systems under investigation (see Tab. VII). All measured data are shown in Fig. 15, including those where the leachates have been mixed intensively before sampling (sampling campaigns before 2001).

In NaCl solutions, the uranium concentrations increased within weeks to a constant level. In $MgCl_2$-rich solutions cement blocks with W/C ratios between 0.30 and 0.34 l/kg showed also from the beginning a constant uranium concentration. In NaCl systems (Fig. 15a), the uranium concentrations were found below $1.0 \cdot 10^{-6}$ mol/l for cement blocks with W/C ratio of 0.5 l/kg. The uranium concentration in $MgCl_2$ systems reached a maximum of $4.0 \cdot 10^{-6}$ mol/l (Fig. 15b). A correlation of the uranium concentration with the sampling position of the solution could not be found. Also, the concentrations of samples filtered by 450 nm filters scattered in the same range as unfiltered samples. The reason for the different release processes from cement blocks of higher and lower W/C ratios could not be explained unambiguously. Reasons could be insufficient mixing during preparation of the cement blocks (blocks #25 to #27 were prepared in 1984, the blocks #31 to #34 in 1986.) Another possibility could be attributed to the precipitation of U bearing solids at the surface which control the solubility.

Tab. VII Fraction of uranium mobilized into the leachate at termination of the experiments.

Block ID	Dopant	Initial Inventory	Fraction of radionuclide found in the leachate %	Duration of the experiments years
#25	$U_{nat.}$	951 g	$(1.1 \pm 0.2) \cdot 10^{-2}$	28.2
#26	$U_{nat.}$	951 g	$(1.1 \pm 0.3) \cdot 10^{-2}$	28.3
#27	$U_{nat.}$	926 g	$(4.5 \pm 1.9) \cdot 10^{-3}$	28.3
#31	$U_{nat.}$	1017 g U / 1370 g $(NH_4)_2U_2O_7$	$(6.6 \pm 1.3) \cdot 10^{-4}$	22.1
#32	$U_{nat.}$	1017 g U / 1370 g $(NH_4)_2U_2O_7$	$(4.2 \pm 3.7) \cdot 10^{-4}$	28.3
#33	$U_{nat.}$	1017 g U / 1370 g $(NH_4)_2U_2O_7$	$(1.6 \pm 0.1) \cdot 10^{-3}$	22.1
#34	$U_{nat.}$	1017 g U / 1370 g $(NH_4)_2U_2O_7$	$(7.7 \pm 3.8) \cdot 10^{-4}$	18.0

By plotting the uranium concentrations versus the pH values of the solutions two trends can be differentiated in the $MgCl_2$ systems (Fig. 16). In the case of the cement blocks with low W/C ratios exposed to $MgCl_2$-rich solution, the pH did not change significantly during the period of the experiments. During the whole period, the pH scattered between 6.1 and 7.0. The U concentration was relatively constant and showed no trend. In the case of the cement blocks with high W/C ratios that were exposed to $MgCl_2$-rich solution, the pH increased during the period of the experiments, affecting the U concentrations. This pH dependence of the U concentrations together with the low amount of mobilized uranium indicated a release process controlled by the solubility of uranium bearing mineral phases. The mechanism will be discussed in chapter 5.2.

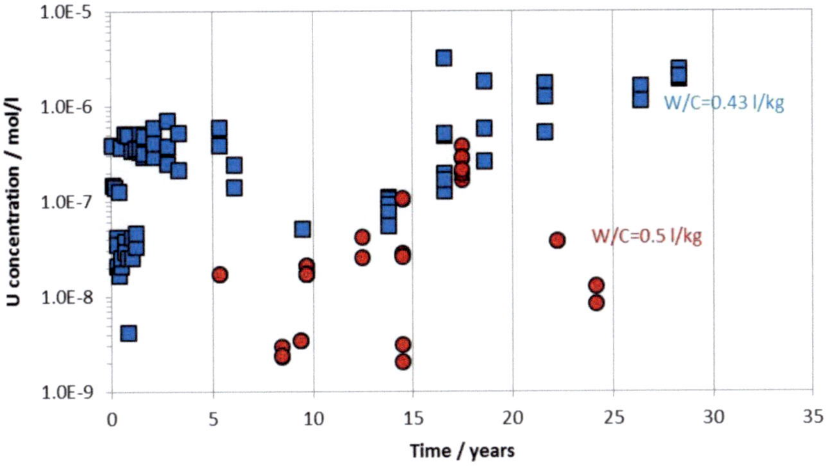

a) NaCl system

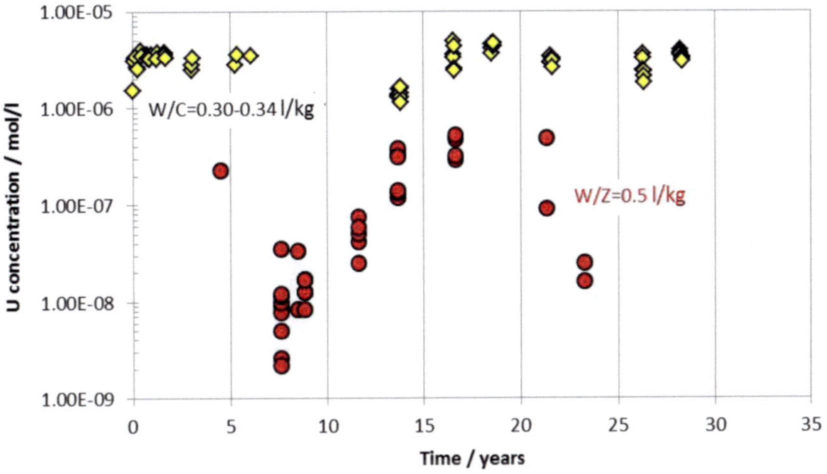

b) $MgCl_2$ systems

Fig. 15 Uranium release as function of time in (a) NaCl and (b) $MgCl_2$ systems, sorted by W/C ratios of the cement blocks. Diamonds: 0.30 ≤ W/C ratio ≤ 0.34 l/kg, squares: W/C ratio = 0.43 l/kg, circles: W/C ratio ≥ 0.43 l/kg.

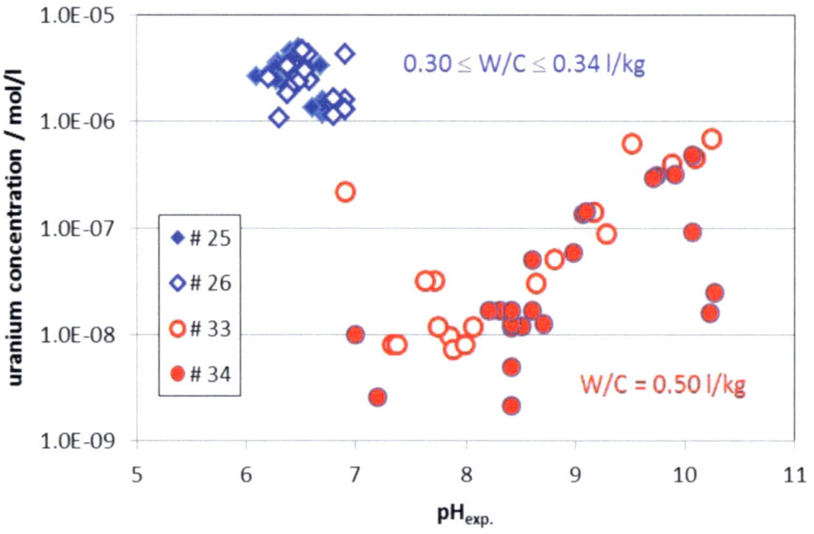

Fig. 16 Measured uranium concentrations as function of the experimentally determined pH of the $MgCl_2$ leachate.

3.2.3.3 Neptunium

The two ^{237}Np doped cement blocks #35 and #36 had a W/C ratio of 0.5 l/kg and an activity of $3.7 \cdot 10^6$ Bq per block (0.145 g ^{237}Np). Block #35 was exposed to NaCl- and #36 to $MgCl_2$-rich solution since 1987.

The ^{237}Np concentration of all solutions of blocks #35 and #36 sampled in May 2001 (i.e. after 11.6 years) was determined below the detection limit of 1.1×10^{-10} mol/l. In 2013 (after 23.3 years of exposure) the detection limit could be lowered to 3.5×10^{-13} mol/l. For this reason, earlier sampled solutions were re-analyzed. The results are shown in Fig. 17. The concentrations measured from filtered solutions indicated constant concentrations. The Np concentrations varied by a factor of 3 between the NaCl and the $MgCl_2$ systems. In both cases, the measured pH and Eh (287 and 273 mV) values indicated the thermodynamic stability field Np(V) [59]. The fraction of released ^{237}Np related to the initial inventory is given in Tab. VIII for the last solution sampling.

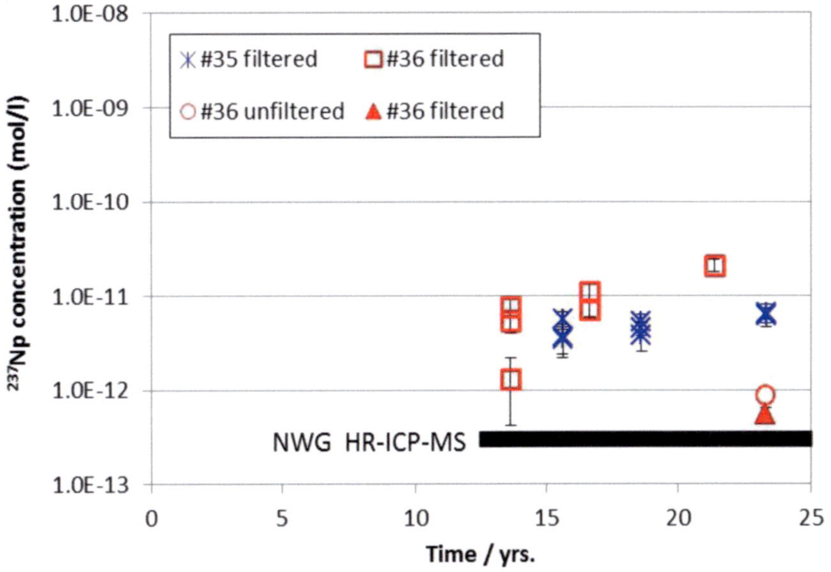

Fig. 17 Evolution of the neptunium concentration in the leachates of cement blocks #35 and #36.

Tab. VIII Released fraction of neptunium from the cement blocks.

Block ID	Dopant	Initial Inventory	Fraction of radionuclide found in the leachate %	Duration of the experiments years
#35	^{237}Np	0.145 g / $3.70 \cdot 10^{6}$ Bq	$< 1 \cdot 10^{-5}$	18.0
#36	^{237}Np	0.145 g /$3.70 \cdot 10^{6}$ Bq	$(5.2 \pm 0.8) \cdot 10^{-5}$	23.3

3.2.3.4 Colloid effect

Some of the leachates showed a cloudy appearance in 2001, 2003, 2006 and 2011. It was assumed that parts of the main components in the solutions were present in colloidal form. To investigate and quantify the colloids, the solutions were filtered by 5 different filter sizes (3, 10, 30, 100 and 300 kDalton). However, the measured concentrations of the main components did not differ significantly after filtration. Filtration of the solutions did not show any effect with respect to the dissolved ^{137}Cs concentrations. The difference found for the solutions from experiment #28 filtered by 3 kDa and unfiltered amounted to less than 2% of the ^{137}Cs activity. Also in the case of uranium and neptunium, no significant differences between filtered and unfiltered solutions were found (Fig. 17).

3.3 Mechanical properties of the solids

The compressive strength of cemented waste forms is relevant in the scope of safety analysis of nuclear waste disposal. Different methods were applied: Elastic properties, such as elastic modulus E, shear modulus G and Poisson's ratio of the solids were determined by application of an acoustic measurement technique [29]. The compressive strength is an exponential function of these parameters [60]. From cores of block #31 cylindrical sections were cut and the ultrasonic frequencies were determined. These could be assigned to flexural and torsional vibrations. The cylinder sections had a diameter of 43 mm, lengths of 103 mm and 77 mm and masses of 277 g and 211 g, respectively. The results are shown in Tab. IX. These data allowed calculation of compressive strength according to equation (2) [60].

$$\beta = a\, e^{bE} \qquad (2)$$

The constant a depends on the composition of the cement and b on the degree of hydration. The compressive strength of hydrated, non-corroded cemented waste products was determined by Vejmelka et al. and were around 30 MPa [5]. This value is slightly below the compressive strength of common hardened cement paste without incorporation of wastes [61]. Vejmelka also reported the compressive strength of cemented intermediate waste products as a function of the $NaNO_3$ load and of W/C ratios [5]. The results are listed in Tab. X, together with measurements made at the full-scale simulated waste blocks.

Tab. IX Mechanical parameters of a drilling core made 15 cm from the cylindrical surface of block #31 (NaCl system).

	Dimension	Depth from top: 40-50 cm	Depth from top: 0-10 cm
Density	g/cm^3	1.89	1.87
E modulus	GPa	0.12 ± 36 %	0.41 ± 36 %
G modulus	GPa	0.06 ± 28 %	0.14 ± 34 %
Poisson number	-	0.3	0.3

The other method mad use of a concrete test hammer (Original Schmidt). Measurements were performed at the top face of the blocks (after removal of loose material) at 15 different spots. An average rebound-value was calculated and by means of a calibration curve, the compressive strength was derived [58]. The average compressive strength of simulates #21 to #26 exposed to $MgCl_2$-rich solution amounted to 18±10 N/mm^2 (Tab. IV). These values corresponded to the initial compressive strength of similar cemented waste forms with W/C ratio of 0.3 and 10 wt.% $NaNO_3$ load. The compressive strength of cemented blocks having higher W/C ratios decreased below the detection limit. In the case of cemented blocks corroded in NaCl solution, the compressive strengths did not change in comparison to the initial values.

Tab. X Compressive strength of the full-scale blocks determined by a rebound test hammer method (Schmidt Hammer).

Block	W/C l/kg	Initial compressive strength at 10 wt.% $NaNO_3$ load [5] N/mm²	NaCl systems compressive strength N/mm²	$MgCl_2$ System compressive strength N/mm²
#21	0.3	20.6		10
#22	0.3			30
#23	0.3		>60	
#24	0.3			28
#25	0.3			10
#26	0.3			<<10
#27	0.4	12.6	10	
#29	0.4			Not measurable
#32	0.5		12	
#34	0.5	7.7		Not measurable
#35	0.5		28	
#36	0.5			Not measurable

3.4 Chemical and mineralogical characteristics of the solids

During the leaching and corrosion experiments some solid samples could be recovered and were analyzed (i.e. from blocks #9, #10, #33, #34 and #36). After termination of the experiments, drilling cores and dust samples of selected blocks were characterized by chemical and petro graphical methods (blocks #14, #20, #25, #28, #30, #31, #32, #33, #34, #35 and #36). The results are summarized and presented in the following chapter, complemented by some particularly interesting data.

In 1989, after 10 years' exposure, solid samples from the inactively doped block #9 (pure hardened cement paste in $MgCl_2$-rich solution) and #10 (in NaCl-rich solution) were taken and a series of thin-section was prepared for analysis by polarization microscopy. Due to the small-sized cement minerals phase identification was not possible, but the texture could be observed. Many fractures were found in the thin-sections. Along fractures of thin-sections from block #10 a change of the color of the cement matrix was observed. Some of the fractures were filled with salt minerals that have been precipitated from the corroding solution. Besides the filled fractures partly filled and finer open fractures were found which appeared to radiate out from open pores. These micro fractures were found in the whole block (see [26]).

X-ray diffraction analyses were performed with materials from cement blocks #9 and #10 resulting in following phases: portlandite, calcite and ettringite (see [26]).

Samples taken from the outer corrosion layers of blocks #33, #34 and #36 and were analyzed by XRD, too. Gypsum, calcite and Mg oxychloride were identified, Gypsum gave highest diffraction peaks. Crystalline C-S-H phases could not be identified.

3.4.1 pH of the pore waters in the solids

Aliquots of all drilling dust samples collected from block #31 and #33 were contacted to distilled water for several days (5 g solid per 25 ml H_2O). The drilling dust of cement block #31 (NaCl system) showed an average experimental pH in the range of 12.5 ± 0.1, drilling dust cement block #33 ($MgCl_2$-rich system) a pH of 12.1 ± 0.2. In both experiments, the pH_{exp} values of the equilibrated solid-water system were slightly higher than of the corroding solutions. It needs to be mentioned that in the equilibrated solid-water systems the ionic strength was significantly lower in comparison to the corroding solutions.

Direct measurements could be performed with liquid collected in the drill hole C (10 cm from the cylindrical surface at 40 cm depth) of the cement block #34. Measured pH of the liquid was in the range of the pH values measured in the corroding leachate: $10.23 \leq pH_{exp}$(drilling dust) ≤ 10.32.

3.4.2 Waste components in the solids

The distribution of the radionuclides cesium, neptunium and uranium as well as the waste component nitrate in the drilling cores / dust samples were investigated. In several of the following graphs data points from drilling dust have assigned horizontal lines (parallel to the x-axis). These indicate the depths from which the individual samples were collected.

3.4.2.1 Cesium

The ^{137}Cs activity distribution in the blocks #28 ($MgCl_2$-rich solution) and #30 (NaCl-rich solution) was determined by γ-spectrometry of the drilling dust. The method was not calibrated, for this reason, the concentrations could not be converted into absolute concentrations for comparison with the overall inventory.

The activity distributions of two analyzed drilling cores from block #30 were found to be similar in all depths > 10 cm. However, the core which was drilled at 6 cm distance to the cylindrical surface showed a slightly lower activity (lower Cs content) in comparison to the core that was drilled closer to the center of the block, at 12 cm distance to the surface. In material from the drillings at 6 cm and 12 cm distance from the circumfence of the block in the depth 0 – 10 cm, the activity was only 50% of the value measured at deeper levels for the coeres. This finding indicates that Cs mobilization occurred mostly close to the surface. In drilling dust samples that were obtained from deeper locations (≥ 12 cm distance towards all surfaces), constant activity was measured, suggesting negligible mobilization of Cs at these levels.

Block #28 had the same dimensions as block #30 and both were doped with the same activity of ^{137}Cs. The activity distributions differed distinctively. In drilling dusts from drillings at 7.5 cm and 13.5 cm from the cylindric surface at depths > 30 cm, the activity was by a factor 3.5 lower than the value determined in the center of block #30. Morerover, an increase of the activity close to the surface was detected in simulate #28. The district reduction of the Cs concentration in $MgCl_2$-rich solution systems (at high W/C ratios) corroborated the high Cs concentrations in the solutions. Higher concentration close to the surfaces may be a result of the association of Cs with newly formed secondary phases. However, inhomogeneous mixture of the cement during the preparation of the blocks could not be ruled out, completely.

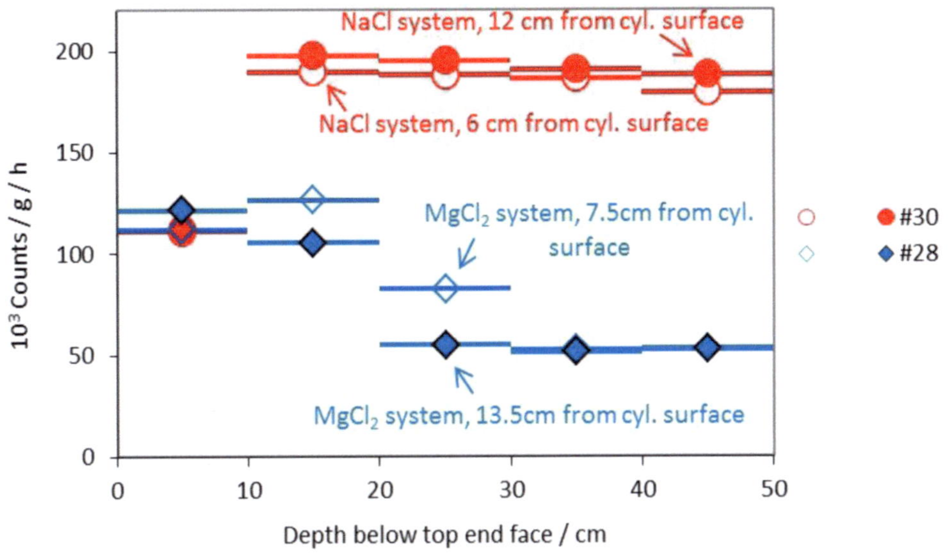

Fig. 18 Relative ^{137}Cs activity distribution as function of the depth in the cement blocks #28 (W/C ratio 0.43 l/kg in $MgCl_2$ solution) and #30 (W/C ratio 0.50 l/kg in NaCl solution).
The distance of the drillings to the cylindrical surface of the cement blocks is shown in the graph (empty symbols represent drillings close to the surface, filled symbols represent more central drilling cores). The γ-spectrometric measurements were performed using equal masses of material and the same measurement geometry.

The distribution of inactive Cs in the blocks was determined in a block which was prepared at W/C ratio of 0.24 l/kg (block #14) and then exposed to tap water instead of brine. The distribution of the remaining Cs is qualitatively similar as in the case of blocks exposed to NaCl solutions. The percentage of residual Cs was determined (Fig. 19). At a depth greater than 4 cm below the surface, Cs was homogeneously distributed and nearly corresponded to the calculated initial concentration of the doped block. However, close to the surface of block #14, the concentration dropped. XRD analyses indicated interactions between tap water and cement only to depths of less than 4 cm ([62]). However, a penetration of water following preferential pathways could not be excluded completely.

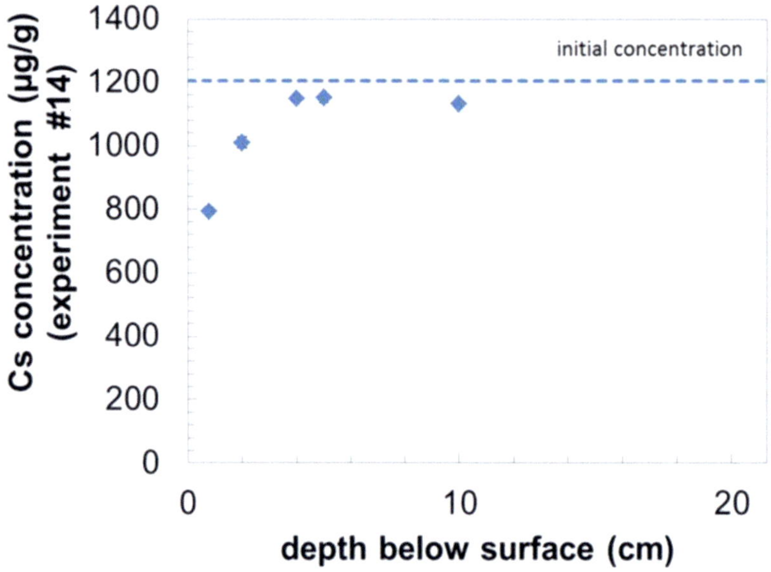

Fig. 19 Profile of the vertical Cs distribution in block #14 after 33 years exposure to tap water.
A vertical drilling core was taken at the center of the upper end face of block #14 (W/C ratio 0.24 l/kg). In this case, solid samples were taken at distinct depths from the core and not from drilling dust. The location of the analyzed samples therefore has a high accuracy in contrast to drilling dust samples.

3.4.2.2 Uranium

For cement blocks #25, #31 and #33, the uranium concentrations in the solids have been analyzed using the drilling dust. For this reason, the concentrations reported in this chapter are averaged over depth intervals of 10 cm, each. It should be kept in mind that the U concentrations in the attacking solutions were low and did not indicate significant mobilization of uranium (chapter 3.2.3.2).

Concentrations of block #25 scattered around the calculated initial concentration (Fig. 20). In the center of the cement blocks #31 and #33 the concentrations were found significantly above the theoretical initial values. Only drilling dust samples obtained close to the end face had a slightly lower U concentration (#31, see Fig. 21) or even a higher concentration (#33, see Fig. 22) in comparison to the calculated average initial concentrations. The measured U profiles in the blocks showed a certain degree of inhomogeneity and moreover, photographs that had been taken before contacting the blocks with brines also indicate inhomogeneous mixing in some parts.

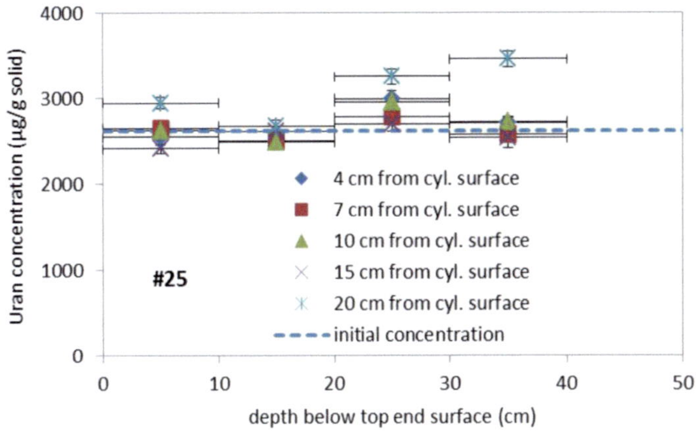

Fig. 20 Absolute uranium concentrations in the solids of block #25 (W/C ratio 0.34 l/kg, exposed to $MgCl_2$-rich solution).

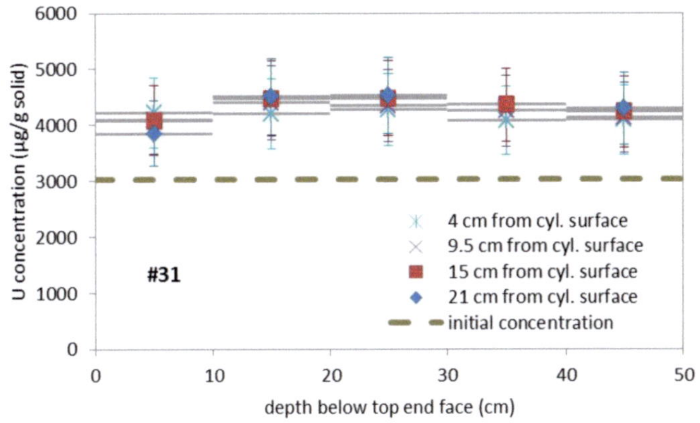

Fig. 21 Absolute uranium concentrations in the solids of block #31 (W/Z ratio 0.50 l/kg exposed to NaCl-rich solution).

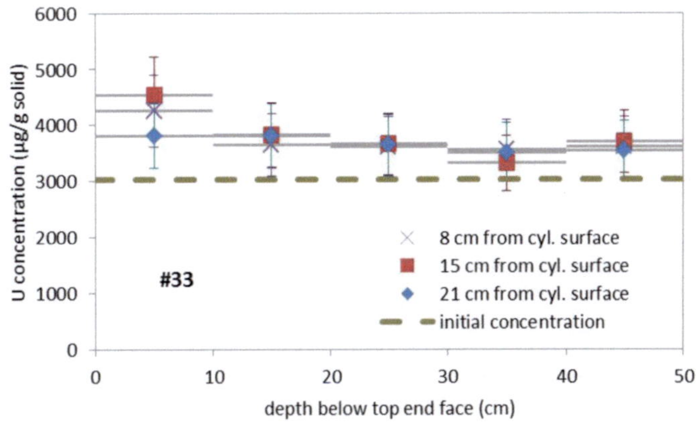

Fig. 22 Absolute uranium concentrations in the solids of block #33 (W/Z ratio 0.50 l/kg exposed to $MgCl_2$-rich solution).

3.4.2.3 Neptunium

In Fig. 23 and Fig. 24 the depth profiles of the neptunium concentrations in the two blocks #35 and #36, respectively, are shown. Both blocks had been produced under the same conditions and were then exposed to NaCl-rich solution (#35) and $MgCl_2$-rich solution (#36). As expected from the low Np concentrations in the leachates (chapter 3.2.3.3), a related reduction of the Np concentration could not be detected in the solids. The concentrations scattered in both blocks around the average concentration of initially doped ^{237}Np. The scatter of the distribution patterns of block #36 was more pronounced (Fig. 24) in comparison to block #35 (Fig. 23). Because of the strong degradation of block #36 in $MgCl_2$-rich brine, it could not be determined whether the scatter of the Np content was caused by the degradation processes or whether it was due to the preparation of the cement block, as in the case of U.

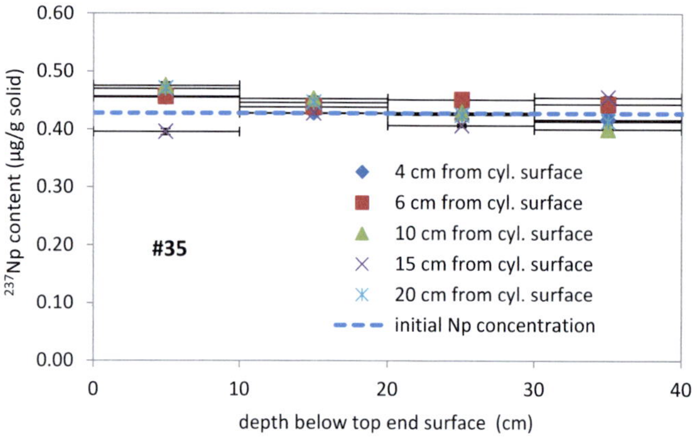

Fig. 23 Absolute neptunium concentrations in the solids of block #35 (W/Z ratio 0.50 l/kg, exposed to NaCl-rich solution).

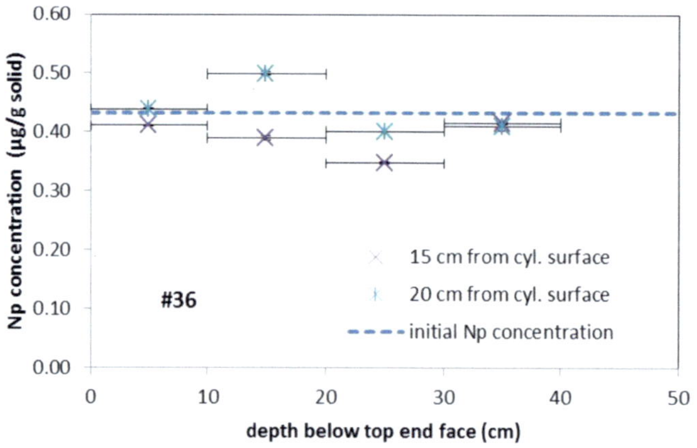

Fig. 24 Absolute neptunium concentrations in the solids of block #36 (W/Z ratio 0.50 l/kg, exposed to $MgCl_2$-rich solution).

3.4.2.4 Nitrate

In the context of Cs mobilization from the cement blocks (chapter 3.4.2.1), the release of nitrate and the dependence on the W/C ratio and brine type was already discussed. The findings of the solution analyses are reflected in the post-mortem-analyses of the cemented blocks. The nitrate distributions in the solids are presented exemplarily by means of selected depth profiles in Fig. 25 and Fig. 26. The drillings were made at FZJ and HDB, using different drill diameters. For this reason, closely adjoining drilling dust samples are compared in the figures.

In the blocks exposed to NaCl-rich solution (W/C ratio 0.43 and 0.50 l/kg) a decrease of the nitrate content was observed from the center towards the surface. The remaining content of NO_3 at depths of 0 – 10 cm was about 46 ± 10 % of the initial amount. The distribution showed a gradient of the concentration over the first 10 cm depth interval. Drilling cores at a distance of > 10 cm from the cylindrical surfaces of the blocks revealed about 100 % of the initial $NaNO_3$ concentration at depths below 10 cm (Fig. 25).

Full-scale simulates that were cemented at high W/C ratios showed nitrate mobilization from all depths in $MgCl_2$ systems (Fig. 26). The NO_3 distribution did not exhibited a gradient in the solids and the average of the remaining nitrate concentrations was 29 ± 11 % of the theoretical initial inventory. The chemical form of the nitrate in the solids could not be clarified. In block #25 $NaNO_3$ crystal bearing lenses were found (chapter 3.4.4.1), but the interior of these lenses probably had no contact to the attacking solution. Similar structures were not found in blocks having higher W/C ratios. Nitrate might form mixed phases by substituting sulfate. However, identification is challenging and little information is available in literature. Decrease of the nitrate content in block #25 (W/C ratio 0.30 l/kg) was found only in depths < 4 cm [31]. This result agrees with the low nitrate concentration in the leachate of this experiment. The higher release of nitrate in NaCl systems where the blocks were produced with the same W/C ratio suggested absence of pore clogging in the NaCl-rich solution.

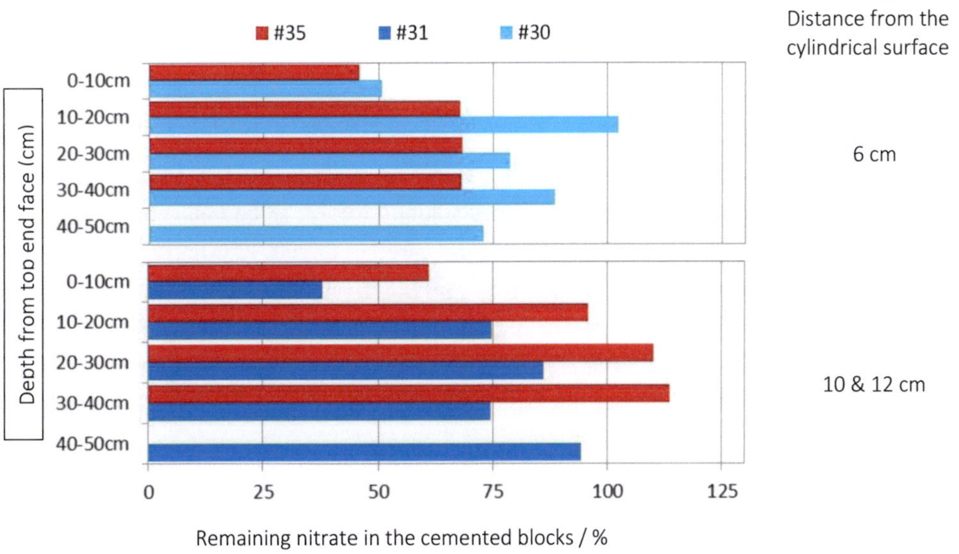

Fig. 25 Percentage of the remaining nitrate content in selected blocks (related to the initial inventory) after exposure to NaCl-rich solution.

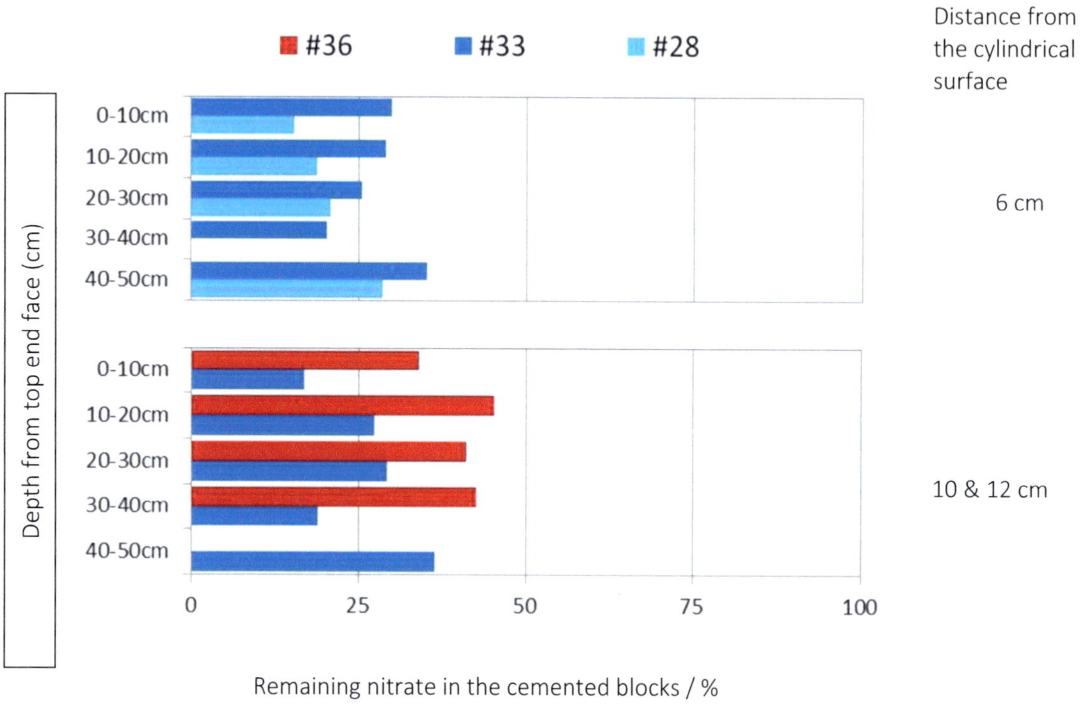

Fig. 26 Percentage of nitrate content remaining in selected blocks (related to the initial inventory) after exposure to $MgCl_2$-rich solution.

3.4.3 Solid phase paragenesis in the solids

Solid phase paragenesis in the solids was determined by XRD analyses using mainly drilling dust. The following XRD patterns show patterns of the mineral phase assemblies of the corroded blocks for several distances of the dust sample from the surface of the blocks.

3.4.3.1 NaCl systems, W/C ratio 0.43 and 0.50 l/kg

A comparison of XRD diffractograms from the blocks of experiments #30 (W/C=0.43 l/kg) , #31 and #35 (W/C=0.5 l/kg) which were corroded in NaCl-rich solution for 22, 18 and 26 years, respectively, is shown in Fig. 27. The results obtained from block #35 are discussed in details: XRD pattern of halite, calcite, Friedel's salt, portlandite and Na-nitrate were found (Fig. 28). Due to the high W/C ratio of 0.50 l/kg, the cement of #35 was completely hydrated and clinker phases were not detected. Strong reflexes of portlandite were present in all samples collected from central drillings as well as from drillings close to the circumference. The presence of portlandite buffered the pH. This was corroborated by the high pH_{exp} of 12.0 determined in the leachate.

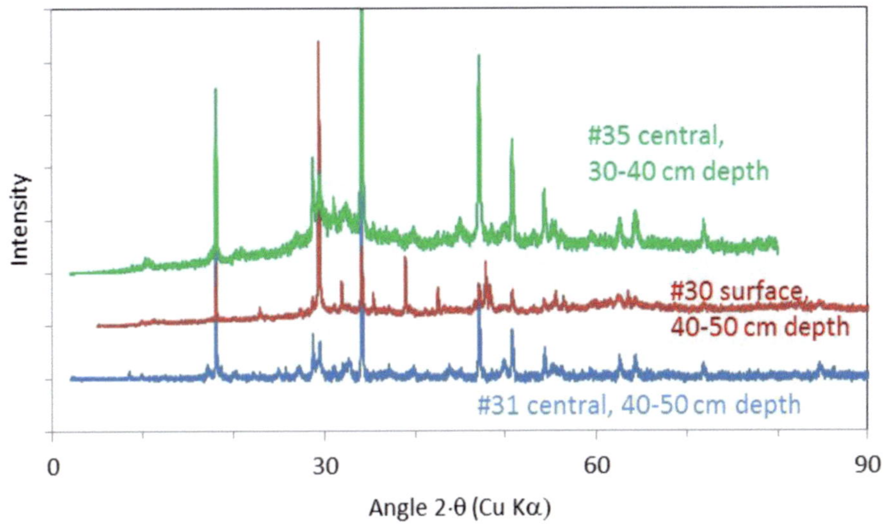

Fig. 27 Comparison of XRD diffractograms from three full-scale cemented waste simulate blocks corroded in NaCl-rich solution.

Halite was present on the surfaces of cement block #35 and was also found to a depth of about 3 - 5 cm into the block. Below 10 cm depth halite reflexes were not recorded in the central drilling core. At the surfaces of the block, Friedel's Salt was present in all samples showing similar intensities of the X-ray reflexes. It was found to a depth of 30 cm and decreasing intensity of corresponding reflections from outside to the central part of the block indicated a decreasing content with depth. In this system, Friedel's salt could be used as a marker for penetration of chloride into the cement paste.

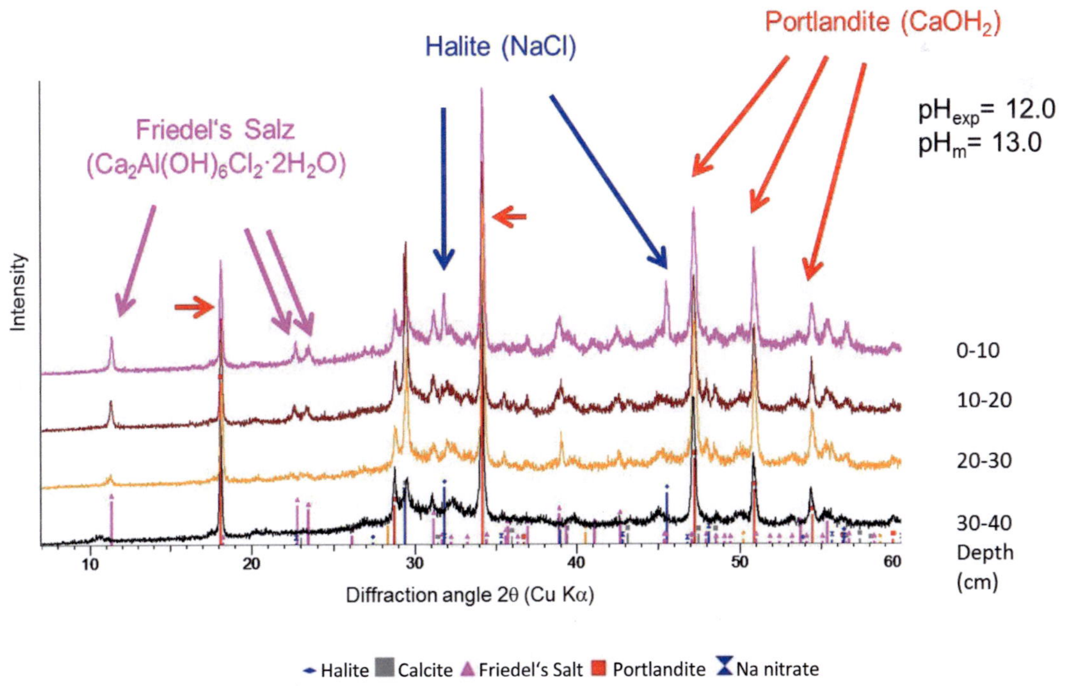

Fig. 28 XRD patterns of drilling dust samples from a drilling close to the cylindrical surface (4 cm distance) of block #35 (NaCl system).

The diffractograms indicated only traces of carbonate phase. As the main diffraction peaks of calcite overlap with Na nitrate and C-S-H-phases, a minor content of calcite could not be ruled out completely. The hardened cement paste of block #35 was only weakly carbonated. In this context, it needs to be stated that this could be the result of the argon atmosphere established in the leaching vessel. Penetration of CO_2 was retarded by the lid system of the leaching container. Carbonate played only an insignificant role for experiment #35.

3.4.3.2 $MgCl_2$-systems, W/C ratio 0.30 l/kg

The block #25 was corroded in $MgCl_2$-rich solution and can be seen as representative for full-scale waste simulates which were cemented at low W/C ratios. All drilling dust samples contained unhydrated cement clinker phases. Ca_3SiO_5 and $Ca_{54}MgAl_2Si_{16}O_{90}$ were identified by XRD. The phase composition of block #25 was characterized by C-S-H, portlandite, cement clinker minerals and the waste component Na-nitrate. Deviations were present only at the deepest position of the central drill hole: A small amount of monosulfoaluminate was identified basing on main XRD peaks at 10.1 and 20.2° 2-Θ. The positions of the peaks were variable to some extend depending on the water content which was affected by the drying process [63] since the amount of water in the interlayers might influence the lattice spacing. Furthermore, in these phases, sulfate could be substituted by nitrate which also affected lattice spacing. At a position close to the surface of block #25, Mg oxychloride was identified as well as weak uranophane reflections. The problem with identifying uranophane unambiguously by XRD is correlated to the fact that this mineral has only one main reflection which overlapped with monosulfoaluminate phases. Additionally, XRD is sensitive for crystalline phases with percentages of at least 1 - 2 wt.% in the solid. The mean uranium content in cement block #25 was at 0.26 wt.%, only.

Also block #25 showed only slight carbonation, only directly on the surface layer chlorartinite was present. Minor amounts of calcite might be present, but could not be identified due to the overlap of the reflections by other phases. The presence of CO_2 from the air played hardly any role in the corrosion of low W/C ratio block. Additionally, the measured pH of the $MgCl_2$-rich solution was around 6 reducing the dissolution of CO_2(g) in comparison with solutions at higher pH and explaining why only little carbonate phases have been formed.

In the surface layer only one typical cement phase (monosulfoaluminate) was identified by XRD, clinker phases and portlandite were not present. Other identified phases were salts, such as halite, sylvite, gypsum, bassanite, Mg oxychloride, syngenite, carnallite, and chlorartinite. Polyhalite could not be identified unambiguously. The precipitation of salt minerals close to the surfaces was controlled by the contact with the $MgCl_2$ leachate, delivering magnesium, chloride and sulfate. Ca was a component of the hardened cement paste and was also mobilized. Sodium and potassium were components of the waste simulate and dissolved from the solid.

At different depths identified solid phases did not reveal any exchange processes between the solid and the leachate below a shallow surface layer. The $MgCl_2$-rich solution had a pH_{exp} of 6 at the end of the experiment, far away from thermodynamic equilibrium with portlandite. One can assume that the kinetics of degradation is retarded distinctly or the corrosion process stopped within the duration of the experiments. The solid below the degradation layer of the cement paste was not in direct contact with the bulk solution.

3.4.3.3 $MgCl_2$ systems, W/C ratio 0.50 l/kg

Inspection of full-scale samples #28, #29, #33, #34 and #36 revealed that the hardened cement paste blocks were extensively degraded. Due to the mechanical properties of these full-scale blocks after the experiments (the material was instable and humid), drilling dust could be sampled and analyzed only from central drillings (Fig. 29).

Drilling dust from 0 – 20 cm depth did not contain any portlandite. At a depth of 20 - 40 cm portlandite was present. The existence of Friedel's Salt and portlandite at the same location indicated a contact of the pore solution with bulk solution. At the high pH in the leachate ettringite was stable and could be identified in all diffractograms. Some drilling dust samples also contained halite. In total, the mineral parageneses, especially the distributions of ettringite and halite, were not homogeneous. A depth depended gradient was not observed.

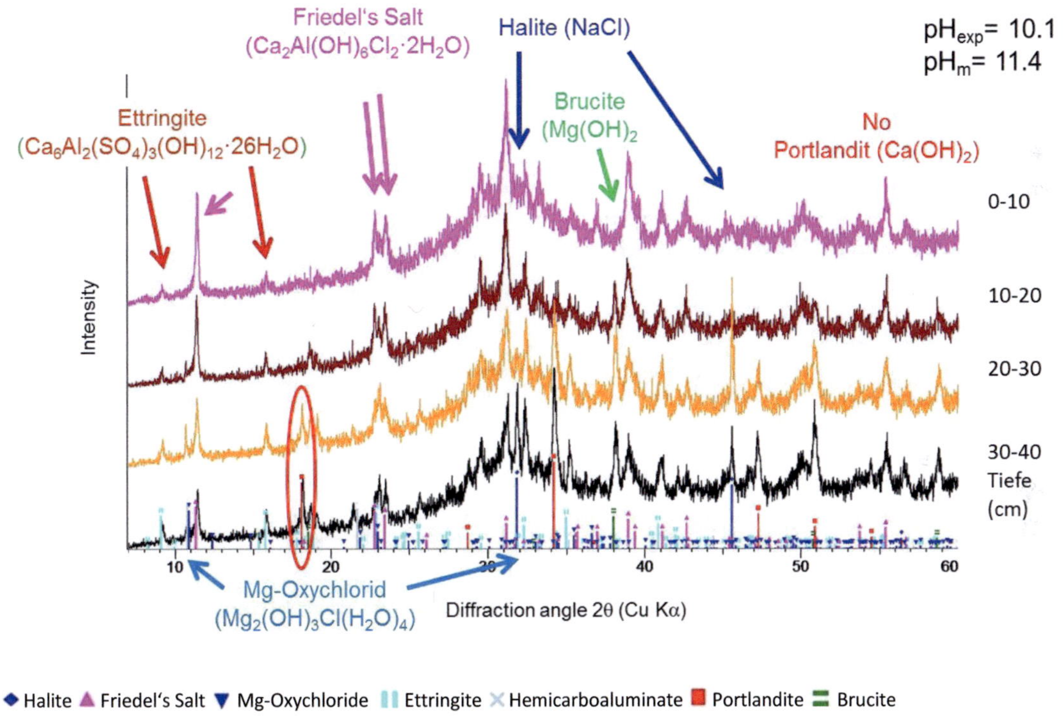

Fig. 29 XRD pattern of a drilling dust sample from a central drill hole (20 cm distance) of the degraded cement block #36 ($MgCl_2$ system).

$CaSO_4{\cdot}nH_2O$ phases were not present in the degraded block #36, but in the similar block #28. Other XRD reflections could be assigned to hemicarboaluminate. The presence of calcite could not be ruled out because of the significant overlap with secondary phases and possibly C-S-H. In any case, the effect of carbonatization to the degradation was negligible in comparison to the massive formation of chloride bearing phases and salts.

3.4.4 Scanning electron microscopic analyses (SEM)

Hints on the mineral phases were also derived by scanning electron microscopic analyses (SEM). By application of an energy dispersive detector for X-rays (EDX), the chemical compositions of phases were determined semi quantitatively. In contrast to XRD, this method does not average over a certain mass of the sample. An indication on new phases formed by the corrosion process could be derived from comparing the elemental compositions of well-known substances with those of the cement corrosion phases, including the structures of the minerals.

3.4.4.1 $MgCl_2$ systems

The following pictures present the appearance and elemental composition of drilling dust samples of full-scale block #28 which was fabricated at W/C ratio of 0.43 l/kg and exposed to $MgCl_2$-rich solution. Mg and Cl bearing phases were found, both Mg and Cl were main components of the corroding solution.

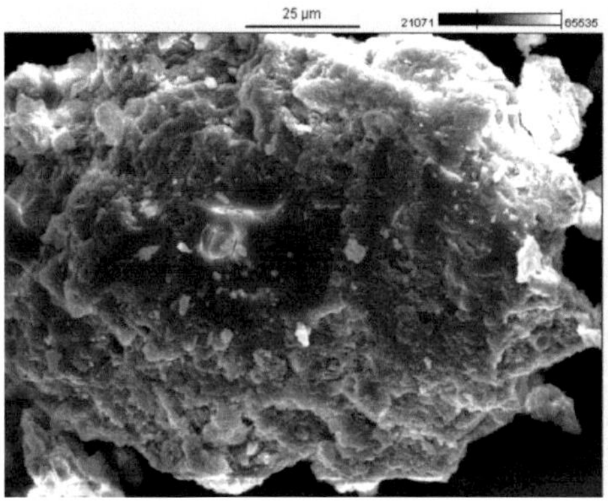

a) Image of an analyzed grain

b) Phase: $Na_9Mg_1Al_4Si_{17}S_3Cl_{18}Ca_{46}$

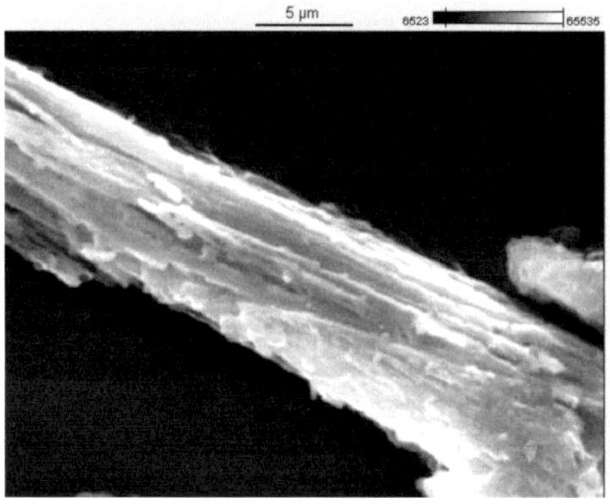

c) Phase: $Mg_{88}Al_1Si_1Cl_4Ca_4$

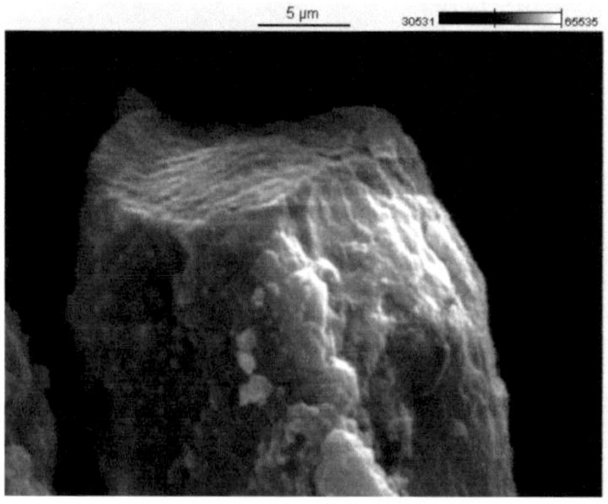

d) Phase: $Na_3Mg_2Al_4Si_{17}S_2Cl_{18}Ca_{52}$

Fig. 30 Encountered phases in the central drilling core of full-scale simulate #28 at 30-40 cm depth below the top face and corresponding semi quantitative results from SEM-EDX analyses.
Oxygen and carbon (used for fixation and coating) are not included into the stoichiometry of the phases because OH groups could not be quantified. As a consequence, the compositions of the phases are not necessarily balanced with respect to the charge.

Microscopic analysis of samples taken from block #25 confirmed the results of the XRD analyses. Cement clinker minerals were present in all distances from the surface, including in layers close to the surface. The $NaNO_3$ phase was not distributed homogeneously over the hardened cement paste, but was present in lenses with diameters up to 1 mm. In these lenses also the clinker minerals were more abundant than in the surrounding matrix. The lenses were delimited from the matrix by a dense border consisting of C-S-H phases. This finding indicates that during curing and hardening of the full-scale block the dense C-S-H layer was formed, preventing further hydration of the inner part of the lenses. A sample appearing optically compact was prepared from a depth of 1 cm below the surface of block #25. Examination by SEM showed a strongly heterogeneous structure. At the microscopic scale, dense and very porous regions were located in close neighborhood (Fig. 31). In the left part of the image a relatively compact region can be seen, whereas in the other parts of the image more porous regions dominate. The same appearance was found over the whole area of analyses in this sample. In both regions, the phase compositions did not deviate significantly. Reaction zones around the pores were not visible. Some single idiomorphic halite crystals of ~ 5 µm diameter were present, surrounded by C-S-H phases.

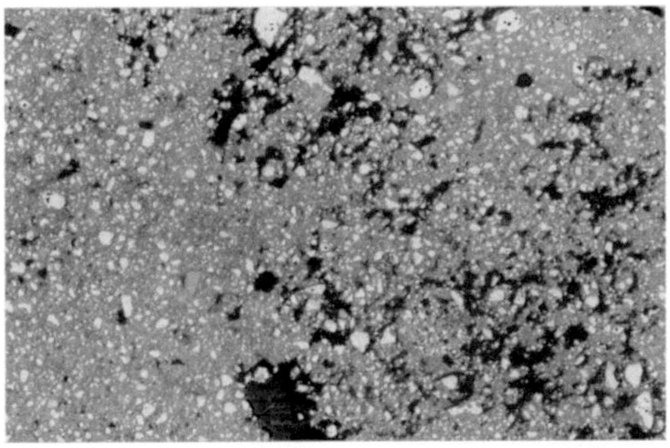

Fig. 31 Backscattering electron microscopic image of the central drilling core in full-scale block #25 at ~ 1 cm depth from the top end face.
The image covers an area of 4.46 x 3.27 mm. Bright spots: mainly tricalcium silicate, pale grey: portlandite, in between C-S-H phases. Dark regions: epoxy resin.

The microscopic analysis of the solid samples taken from block #36 ($MgCl_2$ system, W/C ratio 0.5 l/kg) showed halite crystals in all samples from different depths. Halite crystals had diameters up to several micrometers. The crystals were connected along edges and surfaces and kept an open porosity in the interspaces (Fig. 32). In some regions hydration products of the cement were found, such as C-S-H phases which were degraded to a high extent. The Ca/Si ratio of these phases was in the range of 1. Other regions showed C-S-H phases with Ca/Si ratio up to 2.

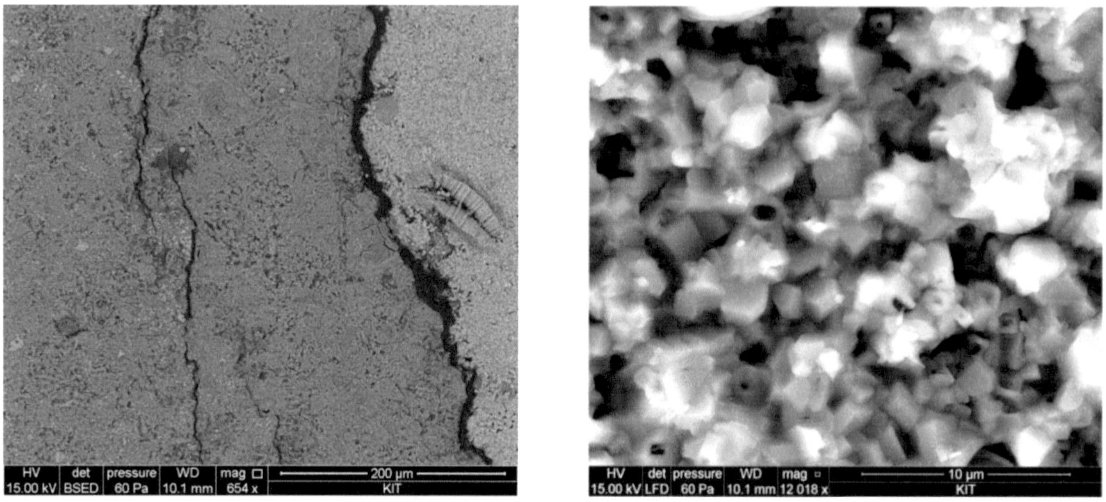

Fig. 32 Backscattering and secondary electron image of a region dominated by NaCl crystals in full-scale block #36 (central drilling).
Left: Backscattering image. The area is dominated by NaCl, but in some places isolated C-S-H phases are present.
Right: Secondary electron image of a NaCl crystal dominated region.

In some images, completely or partly filled pores initially air pores were observed (Fig. 33). These pores are heterogeneously distributed and limited in number. Semi quantitative microanalyses showed always a border of Friedel's salt around the pores. The internal space of the pores was filled with ettringite in many cases. The elemental ratios of aluminum, sulfur and calcium were applied for identification. Sodium and chlorine were present in some of the spot analyses of ettringite, partly at concentrations of a few percent.

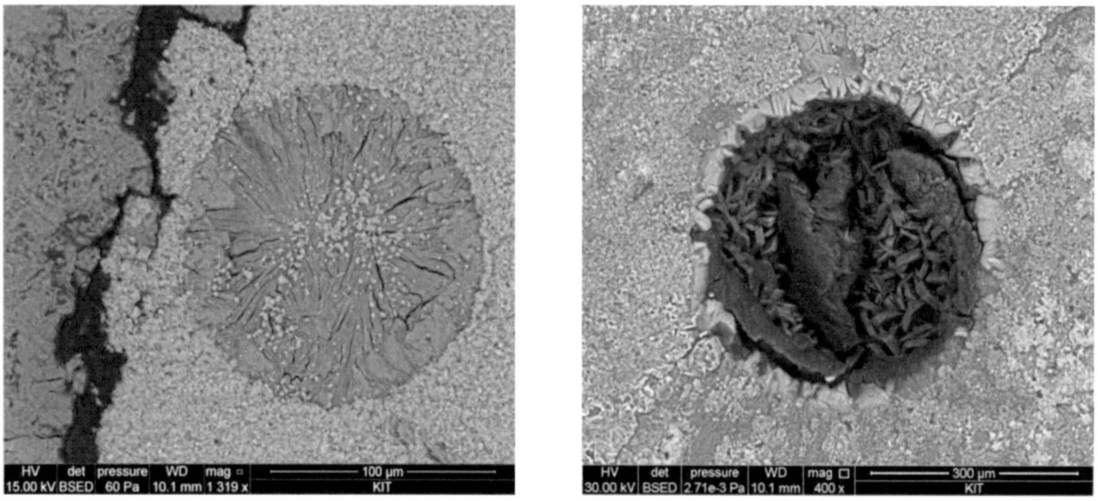

Fig. 33 Backscattering images of a sample from the central drilling core of full-scale block #36.
Left: Pore, filled with secondary ettringite and surrounded by halite crystals. In between a border of Friedel's salt is present. Small bright crystals within the pore are also halite.
Right: An empty pore with a similar border of Friedel's salt, but without ettringite in the center.

3.4.4.2 NaCl systems

Samples from full-scale simulates #31 and #35 which were exposed to NaCl-rich solutions were analyzed by-SEM-EDX also. Fig. 34 shows SEM images of some phases encountered in the peripheral drilling core of block #31.

a.) Elemental composition: 92% Ca
Dominating phase:
Portlantite: $Ca(OH)_2$

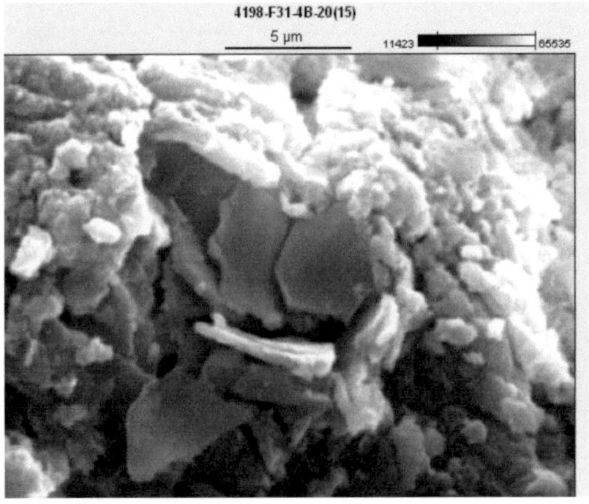

b.) Elemental composition: $Na_4Al_{23}Si_6S_1Cl_{20}Ca_{47}$
Dominating phase:
Friedel's salt: $Al_{24}Cl_{24}Ca_{50}$

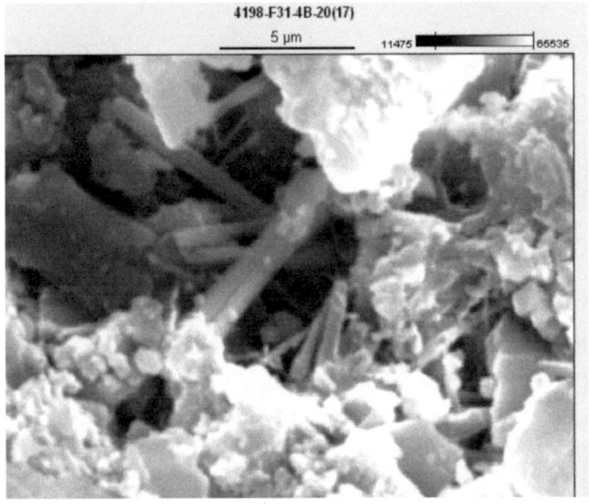

c.) Elemental composition: $Na_{15}Mg_3Al_5S_{25}Cl_7Ca_{42}$
Dominating phases:
Ettringite: $Ca_6Al_2(SO_4)_3(OH)_{12}\cdot 26H_2O$: $Al_{18}S_{28}Ca_{54}$
Monosulfate: $Ca_4Al_2(SO_4)(OH)_{12}\cdot 6H_2O$: $Al_{25}S_{12}Ca_{62}$

Fig. 34 Encountered phases in the peripheral drilling core of block #31 (40 mm from the cylindrical surface) at 10-20 cm depth below the top end face and corresponding semi quantitave results from SEM-EDX analyses
The major elements from semiquantitative analysis are indicated. Oxygen and carbon are excluded. Pure portlandite (a) as well as mixed phases consisting of calcium, aluminum and chloride (b) and sulfate bearing compounds (c) were found.

3.4.4.3 Investigation of uranium phases

In the full-scale cemented block #25 (W/C ratio 0.3 l/kg) uranium phases were found at all sampled positions. The isolated U phases were present in the form of porous aggregates having diameters of about 10 - 50 µm (Fig. 35). However, these phases were found to be unstable under electron irradiation. For this reason, the intensity and accelerator voltage (energy) of the electrons were reduced. For a clear identification of the U phases, single measurements were performed at 30 kV accelerator voltage.

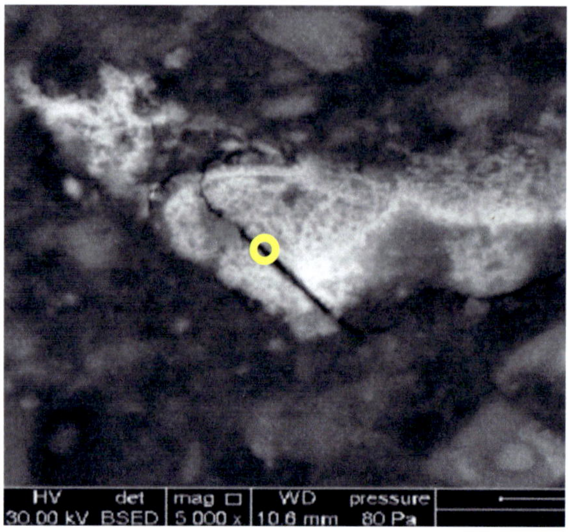

Element	(Atom-%)	
C	32.0	± 2.3
O	43.6	± 2.5
Na	2.6	± 0.4
Mg	0.6	± 0.1
Al	0.9	± 0.09
Si	4.2	± 0.2
P	0.2	± 0.07
S	1.0	± 0.1
Cl	0.8	± 0.1
K	0.4	± 0.1
Ca	11.5	± 0.2
Fe	0.3	± 0.06
U	2.0	± 0.5
Total	100	

Fig. 35 Backscattering image of an uranium aggregate at ~ 1 cm depth below the top end face of the central drilling core in full-scale block #25.
The circle represents the area of analysis.

In some places the uranium aggregates showed a flaky, spiky habitus (Fig. 35), while in other places no clear habitus was recognizable (both types can be observed in Fig. Fig. 36). SEM-EDX analyses of the needles revealed a very similar chemical composition as the other uranium phases and they also were unstable against the electron beam. During analysis of a more compact appearing U phase, the region under investigation disintegrated and the flaky consistency of the phase was visible [31]. From these investigations, one could assume that in the hydrated cement uranium phases appeared generally in flaky spiky consistency. The more compact structures were assumed to be an artifact produced by polishing the samples.

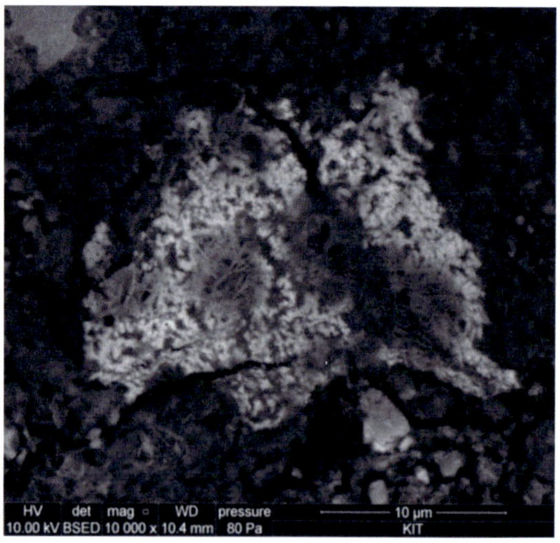

Fig. 36 Enlarged image of an uranium phase at ~ 1 cm depth of the central drilling core in full-scale block #25.
The composition of this U phase aggregate is similar to the one shown in Fig. 35. Crystals of both habitus, flaky, spiky aggregates and more compact crystals are present.

The chemical compositions of the U phases encountered in full-scale block #25 could not be quantified unambiguously due to the instability under electron irradiation and the small diameter of the crystals. A clear differentiation to the Ca and Si dominated cement phases was therefore not possible. EDX analyses revealed a correlation of uranium with phosphorous (the total load of the waste simulate was 4.0 mol uranium and 9.7 mol n-tributylphosphate), but it could not be clarified, if this phase was a phosphor uranyl mineral. However, the presence of chlorine in all uranium phases indicated that U was present in a mixed phase and proved that the U minerals were in contact with the solution.

Also in the full-scale block #31 uranium bearing phases with diameters of 20 - 100 µm (Fig. 37) were detected. These phases were larger than those detected in block #25 which allowed more detailed investigations.

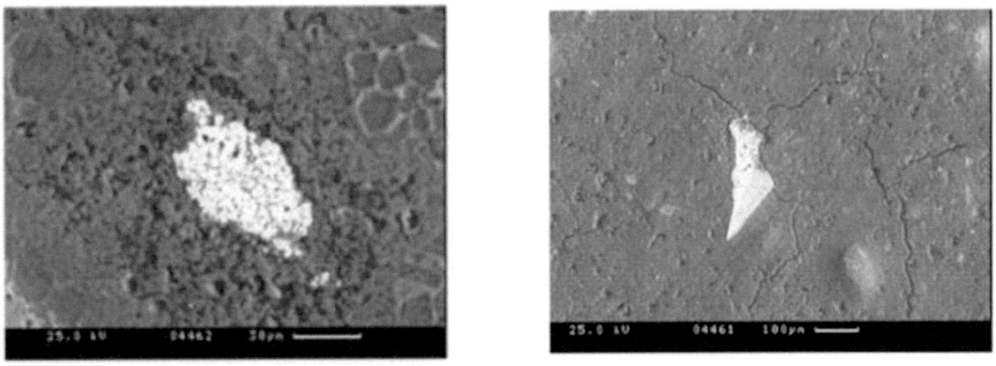

Fig. 37 SEM images of uranium phases in full-scale cement block #31.
The figures give backscattered electron images of uranium phases (bright due to the high atomic number) in the matrix of degraded hardened cement paste.
Scale bar: 100 µm

3.5 Analyses of homogeneously distributed uranium phases

Uranium phases were investigated in samples from full-scale blocks #25, #31 and #33. Detailed descriptions have been published [31, 33, 35, 36]. This chapter gives a short summary.

3.5.1 X-ray diffraction (XRD)

X-ray diffraction patterns of uranium doped samples showed additional reflections in comparison to uranium free samples. Diffractograms of samples from the uranium doped blocks #31 and #33 showed reflections at 23° 2θ (Cu-Kα) (Fig. 38), where the second main reflection of uranophane / β-uranophane is located($Ca(UO_2)_2(SiO_3OH)_2{\cdot}5H_2O$, ICDD powder diffraction files 08-0442, 08-0301 or 039-1360). The main reflection at 11.3° could be identified only in material from block #31 (NaCl-rich solution). In the diffractograms of block #33 ($MgCl_2$-rich solution) reflections of gypsum and Friedel's salt overlap with the the U phase. Additionally to the mentioned reflections, uranophane shows another reflection at 18.2 °, and weaker ones at 34.2° and 48.5° 2θ. However, portlandite and calcite reflect at very similar 2θ angles. For this reason, these diffraction lines could not be used for a clear identification. In Fig. 38 some diffractograms of degraded cement samples are presented together with diffraction lines from the ICDD database, as well as a diffractogram of well-characterized uranophane [64, 65]. Reflections of other uranium bearing phases, such as

$Na_2U_2O_7 \cdot H_2O$ or $CaU_2O_7 \cdot 3H_2O$, meta-schoepite, becquerelite or soddyite, were not detected (Tab. XI). Only uranophane / β uranophane shows significant XRD reflections at 11.3, 23.0 and 31.0 ° 2θ (Cu-Kα).

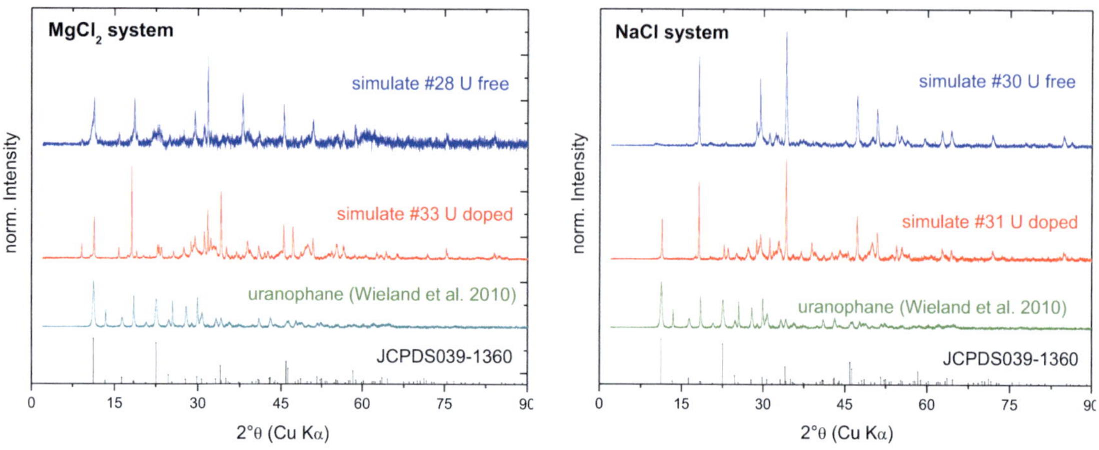

a) $MgCl_2$ systems: #28 (U free) and #33 (U-doped)

b) NaCl systems: #30 (U free) and #31 (U doped)

Fig. 38 XRD diffractograms of cement samples exposed to a) $MgCl_2$ and b) NaCl solution for 17 to 22 years. From top to bottom: uranium free sample, uranium bearing sample, pure uranophane (reference [64, 65]), ICDD database reflections of pure uranophane.

Tab. XI Dominating XRD reflections of relevant uranium(VI) minerals.

Substance	Composition	Positions of main XRD reflections (deg. 2θ) and relative intensities		
Uranophane	$Ca(UO_2)_2(SiO_3OH)_2 \cdot 5H_2O$	11.27 (100%)	23.02 (90%)	31.01 (80%)
β-Uranophane	$Ca(UO_2)_2(SiO_3OH)_2 \cdot 1H_2O$	11.35 (100%)	23.27 (90%)	26.03 (60%)
Sklodowskite	$Mg(UO_2)_2(SiO_3OH)_2 \cdot 1H_2O$	10.54 (100%)	21.57 (80%)	28.10 (70%)
Na Boldwoodite	$Na_{0.7}K_{0.3}(UO_2)_2(SiO_3OH)_2 \cdot 1H_2O$	13.27 (100%)	19.13 (80%)	31.08 (100%)
Meta-schoepite	$UO_3 \cdot 1.5(H_2O)$	12.22 (100%)	17.65 (70%)	26.61 (50%)
Soddyite	$(UO_2)_2SiO_4 \cdot 2(H_2O)$	19.70 (100%)	27.29 (90%)	14.20 (90%)
Ca-Diuranate	$CaU_2O_7 \cdot 3H_2O$	30.34 (100%)	26.89 (100%)	22.65 (100%)

In the powder diffractograms of full-scale block #25 (W/C ratio 0.24 l/kg, $MgCl_2$ system) the reflections at 11.3 °2θ and 23.0 2θ were not present or only very weak and showed overlap with other identified phases. Only in a drilling dust sample from 0 - 10 cm depth weak reflections at 11.3, 23.0 and 31.0° 2θ indicated the presence of uranophane.

3.5.2 Time resolved laser fluorescence spectroscopy (TRLFS)

Over the last years, several studies have been published investigating the fluorescence of uranium phases [64, 66-68]. The U phases show significant differences in their fluorescence properties with respect to the peak positions and to the fluorescence lifetime. For some phases it was not possible to measure fluorescence spectra because of their short lifetime and strong quenching. In the case of mixtures of uranium bearing phases, some single phases with intensive bands may superpose and mask other U phases. TRLFS was applied to drilling dust from blocks #31 and #33, taken from different depths below the surfaces. The samples showed intensive, long-lived fluorescence. For identifying the phases, the fluorescence spectra were compared to reference spectra of meta-schoepite, $Na_2U_2O_7 \cdot H_2O$ and $CaU_2O_7 \cdot 3H_2O$, as well as soddyite and uranophane [64]. A delay of 1 µs and a gate width of 1 ms were applied for all recorded TRLF spectra. The spectra were identical between samples from the blocks and similar to the long-lived fluorescence of uranophane (under identical measurement conditions). The fluorescence emission spectrum of uranophane and spectra of samples from full-scale blocks #31 and #33 are presented in Fig. 39.

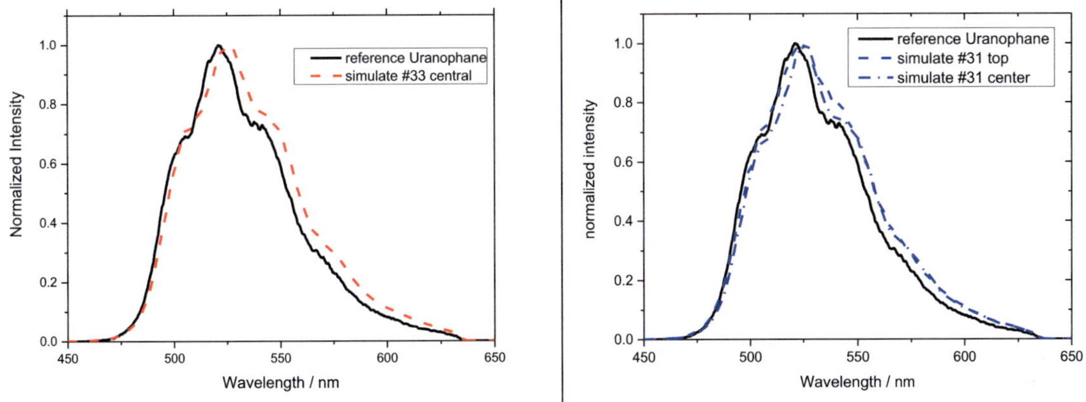

a) Full-scale block #33 ($MgCl_2$-rich solution)

b) Full-scale block #31 (NaCl-rich solution)

Fig. 39 Fluorescence emission spectra of uranium doped drilling dust samples in comparison to uranophane reference material [64].

An example for EXAFS analyses is given in Fig. 40 for simulate #33. The derived structural parameters are listed in Tab. XII. The bond length was in the range of 1.82 Å, corresponding to α -uranophane [69] and to the bond length of the reference compound [64]. The bond length of both axial oxygen atoms was by 0.08 Å larger than in meta schoepite. The coordination number ~5 for the equatorial oxygen atoms coincided with well-known structures of many uranium oxide / hydroxide compounds. However, the bond length of the equatorial oxygen atoms (2.23 Å) was to some extent shorter as expected for α uranophane (2.35 Å, [69]) or 2.28 Å which were measured in the reference compound [64].

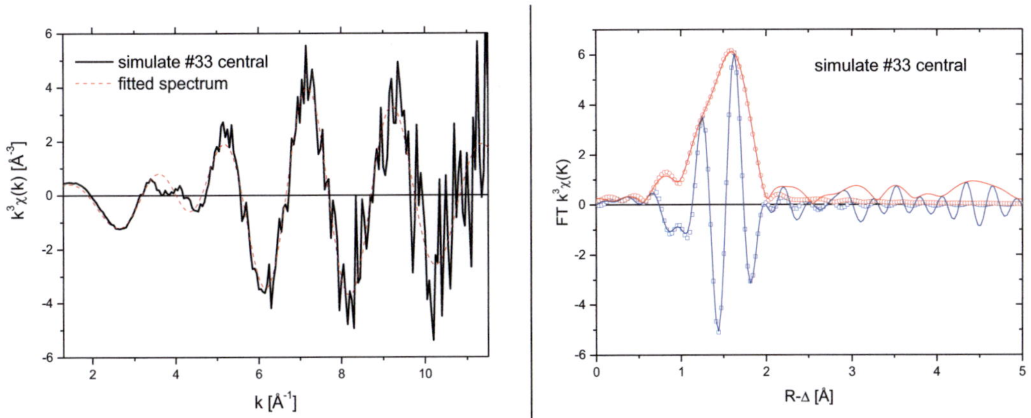

a) EXAFS spectrum

b) Fourier transformation

Fig. 40 U L3 edge EXAFS χ(k) function (right) and corresponding Fourier transformation function (left) of samples from block #33. Measured data: lines, fits: symbols.

Tab. XII Structural parameters of the uranium phase in corroded cement blocks and of a uranophane reference sample [64]. Values fitted from EXAFS measurements.

ID	Coordination shell	Bond distance R(Å)	Coordination number	Debeye-Waller factor σ^2(Å^2)	ΔE_0 (eV)	R-factor
#33-1B-10	O1	1.82(1)	2*	0.003(1)	-9.3(1.3)	0.0003
	O2	2.23(1)	4.7(4)	0.007(1)		
#33-1B-50	O1	1.84(1)	2^f	0.002(1)	-9.5(2.0)	0.0007
	O2	2.23(1)	5.5(7)	0.010(1)		
#31-1B-50	O1	1.85(1)	2^*	0.003(1)	-9.1(3)	0.0001
	O2	2.23(1)	4.0(1)	0.006(1)		
Uranophane [64]	O1	1.82	2	0.002		
	O2	2.28	3	0.005		

Fit parameters: S^2_0 = 1.0, Δk = (1.3 – 11.5) Å^{-1}, Δr = (1.3 – 2.1) Å, *Parameter fixed

3.5.3 Analyses of particulate uranium phases

Particulate uranium compounds detected in the corroded cement blocks are shown in Fig. 35 to Fig. 37 of chapter 3.4.4.3. These compounds were analyzed by µXRD at HASYLAB beamlines at DESY and at the INE beamline at ANKA [33]. The elemental compositions of the U bearing 'hot spots' were determined by µ-XRF. At HASYLAB the µ-XRD diffraction pattern was analyzed by the Laue-Transmission-Modus using a CCD camera; at the INE beamline by means of a X-ray sensitive detector sheet.

Several of the 'hot spots' were analyzed, especially those shown in Fig. 37. These phases had an uranium content of about 1.5 to 2 atom%. The phases turned out to be X-ray amorphous. Weak diffraction lines were assigned to cement corrosion products [33].

Raman spectroscopy was applied to the uranium phases (Fig. 37) taken from corroded full-scale cement blocks #31 and #33, as well as to reference compounds [36]. The spectra are shown in Fig. 41. The comparison indicates that the Raman spectra belong to Na diuranate and not to uranophane. Since Na diuranate was not detected by µ-XRD it must have been amorphous.

As a result of the specific investigations of uranium, it was concluded that at least two different uranium phases exist in the corroded full-scale cement blocks: Uranophane and a diuranate compound.

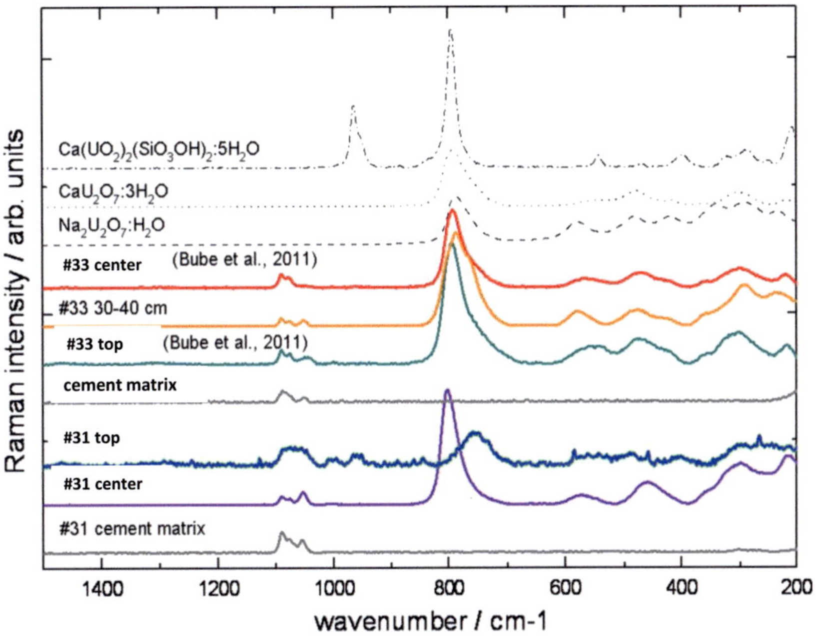

Fig. 41 Raman spectra of different samples taken from full-scale blocks #31 and #33 in comparison with pure uranium phases

4 Supporting experimental results

The full-scale experiments presented in this report have been performed at a constant temperature of 28±1°C which is typical at the depth of the experimental room in the Asse II salt mine, located at 490 m below ground. At the Nuclear Research Center Karlsruhe (today KIT) the leaching and corrosion of simulated full-scale cemented wastes was performed with the simulated blocks KfK #5 and KfK #6 (volume of 160 l, W/C ratio 0.44 l/kg) exposed to 225 l $MgCl_2$-rich solution at 40°C. Also in these experiments, the Ca-Mg exchange was considered as measure of the corrosion progress. Fig. 42 shows the temporal evolution of aqueous Ca and Mg concentrations for the experiments KfK #5 and KfK #6 over a period of 8.5 years. The decrease of the Mg^{2+} concentration and the corresponding increase of Ca^{2+} is clearly shown. After about 7.5 years, the concentrations intercepted. In the same figure, Ca^{2+} and Mg^{2+} concentrations in full-scale experiments #29 (W/C ratio 0.43 l/kg) and #34 (W/C ratio 0.5 l/kg) are plotted. Unfortunately, the evolution of the Ca and Mg concentrations were not measured in the beginning of the experiments. The measured data of #29 and #34 show some scatter, but generally continue the trend from the experiments KfK #5 and #6. At 28°C and 40°C, the Mg concentration is very low after about 8.5 years, indicating that the Ca-Mg exchange reaction was almost complete. For blocks #29 and #34, sulfate concentrations were measured after ~16 and 11 years, respectively, but they were below 0.01 mol/l. Unfortunately, the evolution of the pH of the leachates of blocks KfK #5 and #6 is not available anymore.

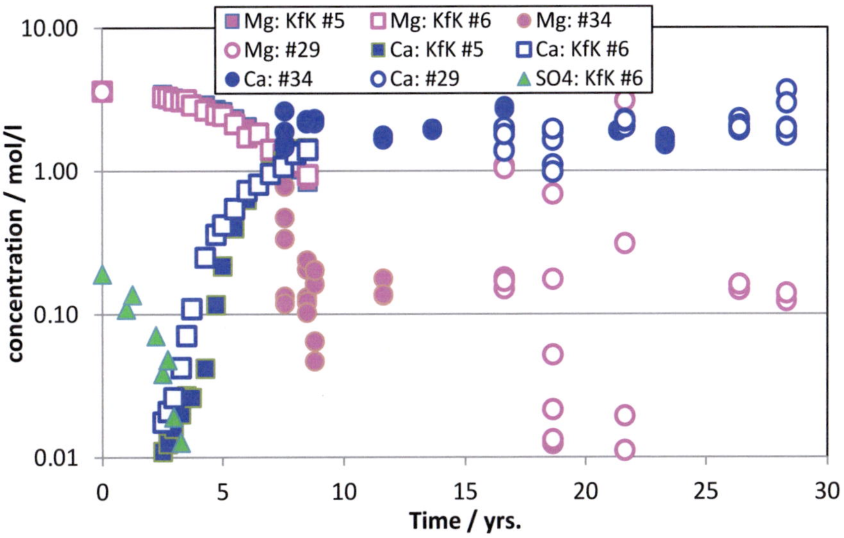

Fig. 42 Comparison of aqueous Ca and Mg concentrations as function of time in experiments #29 and #34 (28±1°C), KfK #5 and KfK #6 (40°C) [24].
The initial Mg concentrations of the leachates were identical. Mg concentrations of KfK #5 almost identical to KfK #6.

In contrast to the Ca-Mg exchange, Cs mobilization was affected by the temperature in the experiments. Cs release to $MgCl_2$-rich solution as a function of time is shown Fig. 43 shows the Cs mobilization as function of time three different types of full-scale cemented waste forms exposed to $MgCl_2$ solution: #28 and #29 are full-scale simulates leached at 28°C, KfK #5 and #6 are full-scale simulates leached at 40°C and AF2 and AF3 are real waste forms also leached at 40°C In this cases, the block volumes amounted to 165 l, stored in

400 l $MgCl_2$-rich solution. Blocks #28, #29, KfK #5 and 6 were prepared with a W/C ratio of 0.43 l/kg and 0.44 l/kg, respectively, the real cemented waste forms were prepared at W/C ratio of 0.65 l/kg.

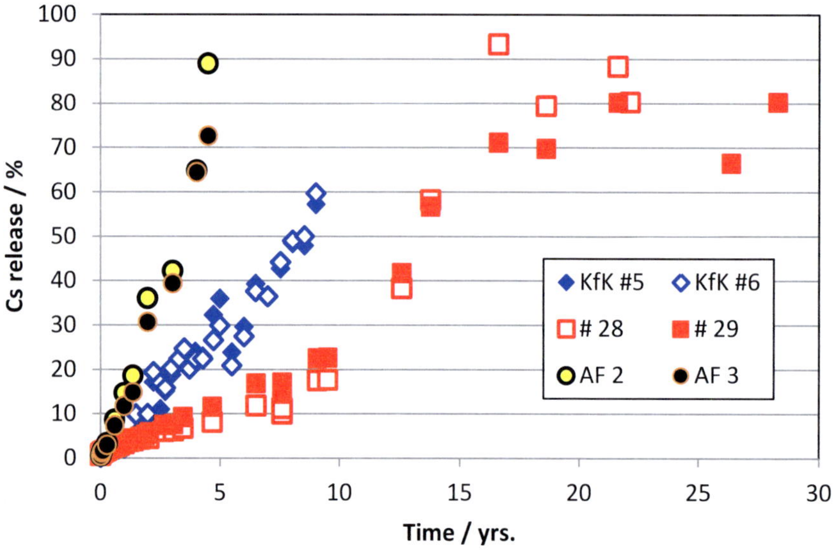

Fig. 43 Cs mobilization as function of time from simulated and real full-scale cemented waste forms exposed to $MgCl_2$ solution.
Red squares: Full-scale experiments #28 and #29 at 28°C, W/C ratio 0.43 l/kg
Blue diamonds: Full-scale experiments KfK #5 and KfK #6 at 40°C, W/C ratio 0.44 l/kg
Circles: Real radioactive cemented waste forms AF2 and AF3 at 40°C, W/C ratio 0.63 l/kg.

The Cs mobilization from simulated blocks stored at 40°C was faster in comparison to the release at 28°C. After 9 years, 60% of the initial Cs content was released at 40°C, a level reached only after ~14 years at 28°C. In the case of the radioactive waste this release was achieved already after 4 years and after 4.5 years, about 90% of the Cs inventory was found in the leachate. This level of Cs mobilization was reached after 15 years for full-scale cement blocks #28 and #29. The real waste forms, however, had a distinctly higher W/C ratio which is known to increase the porosity and therefore facilitate exchange processes.

5 Summary and Modeling

5.1 Cement degradation

The analyses of results obtained from the long-term leaching and corrosion studies, performed at the Asse II salt mine, on simulated cemented evaporator concentrates in full-scale size showed that the W/C ratio adjusted during production of the blocks as well as the type of the attacking solution control the long-term behavior of the cement blocks. The temporal evolution of the full-scale blocks and different achieved degrees of degradation allow for classification into three different groups. These groups can be differentiated by the following experimental parameters:

a) leaching and corrosion of cement blocks having W/C ratios between 0.24 l/kg and 0.50 l/kg in NaCl-rich solution or in tap water,
b) leaching and corrosion of cement blocks having W/C ratios ≤ 0.30 l/kg in $MgCl_2$-rich solution,
c) leaching and corrosion of cement blocks having W/C ratios ≥ 0.43 l/kg in $MgCl_2$-rich solution.

(a) NaCl-rich solution and tap water, W/C ratio 0.24 – 0.50 l/kg

The results concerning the mechanical properties of the degraded cemented waste simulates, summarized in section 3.3, indicate that no measurable decrease in the average compressive strength of the samples occurred after up to 33 years of leaching and corrosion. The measurements by means of a Schmidt hammer were carried out after removing loose material on the surface of the cement blocks. In chapter 3.4.3, it was shown that cement blocks having a W/C ratio ≥ 0.43 l/kg showed reduced compression strengths. The portlandite content close to the surface was reduced and Friedel's salt and halite have precipitated in the pores of the solids. The presence of Friedel's salt in the cement blocks to depths of 20 cm from the surface and unchanged pressure resistance confirm the assumption that the formation of Friedel's salt created no relevant crystallization pressure which could deteriorate the stability of the hardened cement paste. Electron microscopy showed lineament structures. Diffusive penetration of chloride into the cement blocks was detected. It can be assumed that the transport of chloride occurred preferentially along lineaments.

The evolution of the geochemical environment in the leachates showed that in NaCl – hardened cement paste –waste simulate systems equilibrium was established within relatively short periods of time (few years) [35, 70]. Within one year, the pH reached equilibrium. Accordingly, a constant Ca concentration was present as to be expected by the presence of portlandite and the degradation of the cement seemed to have come to a halt. However, until termination of the experiments, slow diffusive transport processes concerning non-solubility limited elements/species still took place, as demonstrated by the penetration of chloride and the mobilization of cesium and nitrate.

(b) $MgCl_2$ solution, W/C ratio 0.24 - 0.34 l/kg

The average compressive strength of degraded cemented waste simulates, prepared at W/C ratios from 0.32 l/kg to 0.34 l/kg and exposed to $MgCl_2$ brine for 32 years did not differ from uncorroded material (chapter 3.3). Three of five simulated cement blocks showed lower compressive strength ≤ 12 N/mm^2 in comparison to laboratory samples showing compressive strengths of 20 N/mm^2 [5]. The mineralogical depth profiles of the cemented blocks showed degradation only for a thin layer of a few centimeters (maximum) below the surface. Below a depth of 4 cm from the surface, no hints of reactions of the cement matrix or the waste components were observed. At the surface salt minerals were found, as well as gypsum, bassanite, chlorartinite and Mg oxychloride. If an internal zoning of the degradation layers existed could not be concluded from the effectuated analyses.

In the case of cement blocks prepared at low W/C ratio ≤ 0.34 l/kg the solution composition was not visibly influenced by the cementitious material until the end of the experiments. Hardly any change of the Ca or Mg concentrations were observed, as well as for the pH. Hence, only limited Ca-Mg exchange between solids and leachates may have taken place. Portlandite and C-S-H-phases at depths larger than ~ 4 cm were not in contact with the leachates, having a slightly acid $pH_{exp.}$ of about 6 to 6.5. Also the spatial profiles of nitrate and chloride in the solids did not show penetration to depths greater than ~ 4 cm. One can assume that the precipitation of secondary phases, especially brucite and gypsum, in the pores prevented further penetration. The corresponding simulates did not follow the evolution that would be expected if assuming thermodynamic equilibrium. XRD analyses of the surface layer directly in contact with the $MgCl_2$ brine of block #25 resulted in newly formed secondary phases (especially Friedel's salt, gypsum, Mg oxychloride and other salts) in equilibrium with the solution. These phases were not present in at greater depth in the cemented block. It is reasonable to assuming a passivating and corrosion-inhibiting layer at the surface, partly filling the pore space and that in the following prevented exchange processes with deeper layers of the cement block.

<u>(c) $MgCl_2$-rich solution, W/C ratio 0.43 - 0.50 l/kg</u>

In contrast to the previously described evolutions of the full-scale cemented blocks, the experiments in $MgCl_2$ brine with cement blocks prepared at W/C ratio ≥ 0.43 l/kg showed strong degradation of the cementitious matrix. Compressive strength was not measurable. This observation is reflected by the XRD and SEM analyses. The results show only a minor amount of portlandite which is inhomogeneously distributed in the solids. Secondary phases such as Friedel's salt, brucite, Mg oxychloride and halite were found in all drilling dust/drilling core samples. Sample preparation and SEM –analyses revealed a porous structure of the solids.

The evolution of pH and the aqueous Ca and Mg concentrations showed a prominent dependence of the degradation kinetics on the initial W/C ratio. In contrast to systems prepared at W/C ratio 0.43 l/kg which achieved equilibrium after about 30 years, the Ca-Mg exchange was completed already after 15 years for cement products prepared at W/C ratio 0.5 l/kg. The pH buffering $MgCl_2$-rich solution together with the volume changes and partly associated crystallization pressure due to formation of secondary Mg and sulfate phases characterize the degradation. As the initial W/C ratio influences the absolute porosity and the pore structure, it is reasonable to assume an effect of the W/C ratio on the degradation kinetics. The threshold value below which the degradation was prevented, probably by the formation of a 'passivating' layer can be estimated in the W/C ratio range between 0.34 and 0.43 l/kg.

Comparison with thermodynamic calculations [35, 70] indicates that the degradation process of the full-scale cement blocks prepared at high W/C ratio that were exposed to $MgCl_2$-rich solution came to a stop because a state of equilibrium was achieved.

5.2 Solubility of uranium(VI) and neptunium(V)

Under oxidizing or redox-neutral conditions, dissolved uranium exists in the hexavalent state in the form of UO_2^{2+} hydrolysis complexes [71, 72]. Solubility and other thermodynamic data were reviewed by OECD-NEA [71, 72]. The solubility of U(VI) in concentrated $MgCl_2$-rich solutions was investigated experimentally by KIT-INE [73] over a pH range from slightly acidic to alkaline conditions limited by precipitation of $Mg(OH)_2$ or Mg oxyhydroxides. It was found that the U(VI) concentration in solution was controlled by meta schoepite ($UO_3{\cdot}2H_2O(cr)$). The solubility dependeds strongly on the ionic strength. In the full-scale experiment #25, the aqueous uranium concentration was slightly below the value determined for the pure $MgCl_2$ solutions. These findings indicate that the uranium concentrations in the solutions were controlled by solubility of a solid phase. Sorption processes played an insignificant role under the conditions of the full-scale experiments.

The differences in U concentrations in the NaCl- and in den $MgCl_2$-rich solutions can be explained by ionic strength effects. Maximum measured uranium concentrations in the two types of leachates were:

- $[U]_{max}$(NaCl system) = $2x10^{-6}$ mol/l ($pH_{exp.}$ ~12.0)
- $[U]_{max}$($MgCl_2$ system) = $5x10^{-6}$ mol/l at W/C = 0.30 to 0.34 l/kg ($pH_{exp.}$ ~6.5)
  $5x10^{-7}$ mol/l at W/C = 0.50 l/kg ($pH_{exp.}$ ~10.0).

Correlating the uranium concentrations determined in the various leachates with the pH values of the $MgCl_2$ leachates resulted in significantly lower uranium concentrations as expected if meta schoepite would control the solubility. This indicates that a phase having lower solubility than meta schoepite controlled the uranium concentrations in the leachates of the full-scale experiments.

The investigations demonstrate clearly that the uranium mobilization is controlled by solids in both solution types. The solid phases are independent of the W/C ratio. In experiments where pH was constant, uranium concentration remained also constant. In the $MgCl_2$-rich solution experiments, where the pH increased due to the dissolution of hydroxide liberating Ca phases from the hardened cement paste (as consequence of the Ca-Mg exchange process), the uranium concentration increased until the solubility limit was achieved. The co-existence of diuranates and uranophane was shown by different applied analytical techniques. However, the existence of other U phases such as U phosphate phases could not be ruled out completely. The existence of diuranate phases in form of relatively big particles (> 50 µm) and the shape of the particles allow the conclusion that the accumulation process did not occur by diffusion processes during the exposure period, but probably happened already during preparation of the blocks. It is assumed that the particles were formed spontaneously when the acidic, uranium containing solution came into contact with cement, changing from a more acid to an alkaline regime.

Under oxidizing or redox-neutral conditions, dissolved neptunium exists as pentavalent cation [74]. Thermodynamic data have been reviewed in the frame of the OECD-NEA 'Thermodynamic Database Project' [72, 75]. The solubility of Np(V) was measured in salt solutions such as 5 molal NaCl solution [76]. Estimation of Np(V) solubility in in $MgCl_2$ solution mostly bases on data for NaCl solution. In literature only one conference contribution exists that presents experimental solubility data of Np in $MgCl_2$ solution [77].

In a 4 molal $MgCl_2$ system at a pH_m range between 8 and 9, the solubility of Np(V) is in the range of 10^{-3} - 10^{-2} mol/kg [77]. For full-scale block #36 the initial composition of the leachate has changed by the corrosion processes (Fig. 11). At the end of the experiment, the measured magnesium concentration was below 0.02 mol/l, while the calcium concentration reached 1.79 mol/l and the sodium concentration 2.58 mol/l. Hence, the behavior of Np(V) was not primarily controlled by the pure $MgCl_2$ system the experiment was started with. Investigations to determine the Np(V) solubility in $CaCl_2$ solutions at different ionic strengths are ongoing at KIT-INE [78]. The experiments show precipitation of $Ca_{0.5}NpO_2(OH)_2{\cdot}1.3H_2O(s)$ and control of solubility by this phase in pure, 2 molar $CaCl_2$ solutions. The Np concentration was in the range of 10^{-8} mol/l in pure $CaCl_2$ solution. This concentration is by four orders of magnitude above measured concentrations in the leachate of block #36. If the Np inventory was mobilized entirely from the cement matrix, the concentration in solution had exceeded the cited solubility limitby two orders of magnitude. These observations and considerations support the assumption that the retention of Np(V) is controlled by sorption processes.

It was shown experimentally that in NaCl solutions Np(V) precipitates in the form of $NpO_2OH_{(am)}$ which then is transformed slowly to the more stable $NpO_2OH_{(aged)}$. The solubility of this phase is by half an order of magnitude lower than the solubility of the initially formed amorphous phase [79]. The Np(V) concentration, controlled by solubility of $NpO_2OH_{(aged)}$, is in the range of 10^{-5} mol/l at pH_m 13.0 [78]. This pH was measured in the full-scale experiment #35 in ~ 5 molar NaCl-rich solution. However, the measured Np(V) concentration was about 5 orders of magnitude below this value. This is a strong indication on the retention process which is assumed to be controlled by sorption and not by solubility of solid Np phases.

5.3 Modeling of cement degradation

The software 'The Geochemist's Workbench' (GWB) [80] was applied as tool for thermodynamic modeling of the investigated systems. The code can handle Pitzer parameters so that the highly concentrated systems can be modelled with a suited chemical activity model.

The evolution of the cement-salt solution systems was simulated as a function of the ratio between cement mass m and solution V. Modeling results were compared to the experimental findings of the full-scale experiments. Together with results from additional laboratory experiments a wide range of m/V ratios was included in the comparison [70]. With the GWB module „React" the initial composition of hydrated cement paste was defined by the relevant element oxides (CaO, SiO_2, MgO, etc.) which react instantaneously with the solution at the appropriate W/C ratio. The salt solutions introduced in the model accordingly to their measured initial compositions before the start of the experiments. The solid to solution ratios were increased stepwise. The stoichiometric composition of the oxides represented cement clinker type *CEM I 32.5 R* (Portland cement). This approach was chosen to follow the stable solid phase parageneses and the according solution compositions which allows direct comparison with the experimental systems. $NaNO_3$, the main component of the waste simulate, was considered in the model, too.

The high ionic strengths of the solutions required application of the Pitzer model for describing the chemical activities [81]. The thermodynamic data was taken from Harvie, Møller and Weare [82] for the oceanic salts, complemented by aquatic species of Si [83] and Al [84]. The relevant Pitzer parameters were taken from [85] and [86]. Relevant solid phases were selected from several databases [83, 87-91]. In the case of solid solutions, ideal miscibility was supposed and the different end members were defined. The sequence of miscibility of the C-S-H solid-solutions was modeled according to [90] applying end members of different Ca/Si ratios. The thermodynamic data were consistently compiled in a database for the calculations.

For details of the numeric simulation as well as the modeling results, it is referred to Bube et al. [70]. The calculated results agree well with the experimentally determined data both for full-scale block #32 (NaCl solution) and #33 ($MgCl_2$ solution). The agreement refers to the solution concentrations and the cement hydrates, as well as secondary phases. It was shown that the initial inhomogeneity at the surfaces of the simulate blocks did not affect the agreement between experimental and numerical data. The good agreement between measured and computed results for NaCl and $MgCl_2$ systems supports the assumption that the finally measured constant concentrations of the leachates represented a state of geochemical equilibrium and were not controlled by kinetic effects.

5.4 Modeling of radionuclide mobilization / retention

Uranium(VI) solubility was calculated using GWB. The thermodynamic data and Pitzer parameters for ionic strength corrections were used as described in chapter 5.3. For modeling the behavior of uranium the solubility and complex stability constants of relevant U(VI) solids and aqueous species were taken from Guillaumont et al. [92] and from Altmaier and Neck [93, 94]. Pitzer parameters of U(VI) species originate from Pitzer [95] and Plyasunov et al. [96] as well as from Neck et al. [94] who derived a parameter set from a correlation of SIT and Pitzer parameters following an approach of Grenthe and Plyasunova [97] and Plyasunov et al. [98]. Pitzer parameters for interactions of Ca^{2+} with U(VI) aqueous species are not available. Results were obtained for a series of uranium minerals such as becquerelite ($CaU_6O_{19}\cdot 11H_2O$), meta schoepite ($UO_3\cdot 2H_2O$) and soddyite ($(UO_2)_2SiO_4\cdot 2H_2O$). The calculated aqueous U(VI) concentrations in equilibrium with these phases were by more than two orders of magnitude above the measured U(VI) concentrations in the full-scale experiments.

Due to the high detection limit of Si (around 10^{-3} mol/l) in the highly concentrated solutions, the calculated solubility of uranophane ($Ca(UO_2)_2(SiO_3OH)_2\cdot 5H_2O)$) could not be compared directly with the experimental systems. Referring to the geochemical calculations for the leachates the SiO_2 concentration was expected in the range of $5\cdot 10^{-6}$ to 10^{-3} m in the cement - salt brine systems [70]. Assuming this range of Si concentrations, measured U concentrations in the $MgCl_2$ systems may agree well with the calculated uranophane solubility. However, it has to be kept in mind that the Pitzer-data for silicate species are based on approximations [99, 100], as experimental data for Si speciation in the highly concentrated solutions were not available.

In NaCl dominated systems, measured Ca concentrations were substantially lower in comparison to the $MgCl_2$ systems. In this case, calculated uranophane solubility would result in an uranium concentration by a factor of 10 above the experimentally observed value. Solubility calculations with Ca diuranate similarly exceed the measured U concentration by one order of magnitude. In contrast to the $MgCl_2$ system thermodynamic, calculations of the U solubity agree well for Na diuranate [35] indicate that this phase is the U controlling phase.

Referring to the mobilization of cesium, mainly diffusion controlled mobilization was assumed and diffusion parameters calculated by means of a three-dimensional model (cylinder geometry), realized using the FlexPDE-finite element software (PDE Solutions Inc.). Calculations covered Cs^+, Cl^- and NO_3^-. The mobilization was calculated for full-scale simulate blocks prepared at high W/C ratio which were exposed to NaCl solution. The model was not applicable to $MgCl_2$ systems, due to the strong mechanical effect of degradation which did not allow for considering an undisturbed cement matrix.

The diffusion coefficients were fitted to the measured data as a function of time. Due to the availability of experimental data for Cs, the temporal evolution of the leachate concentration was applied, while for Cl^- and NO_3^-, the measured profiles in the solid were used. Details of the model have been published [101]. Apparent diffusion coefficients were fitted as follows: Cl^-: $2.3 \cdot 10^{-11}$ m^2 s^{-1}, NO_3^-: $4.6 \cdot 10^{-12}$ - $10.0 \cdot 10^{-12}$ m^2 s^{-1}, and Cs^+: $1.2 \cdot 10^{-12}$ m^2 s^{-1}. The modeling approach allowed to derive a sorption coefficient for cesium in the corroded cement matrix under the assumption that anions are not sorbed. The obtained value was $K_d(Cs^+)$ = 0.23 ml g^{-1} which was in the expected range for Cs sorption onto cement [102].

6 Conclusions

The observations and findings of the long-term leaching and corrosion experiments using full-scale blocks of simulated, cemented waste provide a huge dataset and support understanding of the behavior of this waste type in contact with salt solutions. The information was used in various safety analyses, such as in the preliminary safety analysis Gorleben [38]. Especially the results obtained from cemented products, corroded in NaCl-rich solution are relevant for other disposal sites in rock salt (e.g. Waste Isolation Pilot Plant WIPP, Carlsbad, MN, USA). The presence of cement products may change the solution composition and pH and thus have an essential impact on the radionuclide release or retention.

The results from this research program, lasting from 1978 to 2013 lead to a better mechanistical understanding regarding the following topics:

- Durability of cemented waste forms.
- Kinetics of the corrosion processes.
- Transferability of data from laboratory tests (small-scale, hand-made) to real waste forms.
- Influence of the production process, especially the W/C ratio, on the leaching and corrosion behavior, as well as the radionuclide mobility.
- Geochemical evolution of the system salt solution - cemented waste forms.
- Radionuclide retention.

The durability of cemented waste forms depends on the porosity (W/C ratio) and the type of attacking solution. $MgCl_2$-rich solution is significantly more detrimental than NaCl-rich solution. However, a certain protective effect, probably by precipitation of Mg hydroxides and Ca sulfates, might delay the progress of corrosion in the case of little permeable cement products. The delay of corrosion in Mg containing brines applies also to the penetration of chloride and the leaching of radionuclides and other waste components. Cement products with realistic W/C ratios (> 0.4 l/kg) corrode within about 15 years. During this time geochemical equilibrium between the solid and the attacking solution is established. The discrete composition of the solution (ion concentrations, as well as pH) and the solid paragenesis (primary cement hydrates and secondary phases) in the equilibrated systems strongly depend on the ratio of cement mass and to the solution volume.

In the case of the contact to NaCl-rich solutions, the reactions between solids and solutions are scarcely affected by the porosity of the cementitious material. The major indicator of cement degradation in solution, pH, rises in NaCl-rich solution within months to a constant value (portlandite buffering) and does not change in the long-term. The composition of the solution and the solid paragenesis in the equilibrated systems is relatively insensitive for cement mass to solution volume changes (in the assessed range).

A measure of the transferability of data obtained from laboratory samples to real waste forms regarding the behaviour of radionuclides is difficult to define. One possibility is to make use of the diffusion constants for cesium and compare the results obtained with laboratory samples and with real waste forms. Modeling revealed a diffusion constant of cesium in NaCl systems which was by a factor of about 5 smaller for laboratory samples in comparison to the full-scale waste forms under the same temperature and W/C conditions.

The influence of the manufacturing process on the properties of cement products (water / cement ratio and the related porosity of the cement products) has a significant impact on their behaviour. This is demonstrated by the Cs release from cement blocks of different W/C ratios in $MgCl_2$-rich solution. In principle, cement hardly retains cesium in contact with solutions. The same is true for the waste component nitrate. The Cs-release to NaCl solution is largely independent of the W/C ratio. Sorption of Cs onto more or less corroded cement phases is minor. Cs release to NaCl-rich solution is significantly faster compared with Cs release to $MgCl_2$-rich solution from full-scale blocks prepared at low W/C ratio. The reason is clogging of the cement products surface by Mg precipitates. In contrast, from full-scale blocks prepared at high W/C ratio, a release of almost the entire cesium inventory occurs in $MgCl_2$-rich solution.

Uranium (VI) is incorporated into solid mineral phases and the uranium concentrations in the attacking solutions are limited by the solubility of the solids diuranate or uranophane. This applies to both solutions studied. The measured uranium concentrations are of the same order of magnitude for both solutions. The co-existence of diuranates and uranophane was shown. The existence of relatively big diuranate particles (> 50 µm) was explained also by the preparation process of the simulate blocks: The uranium containing solution was adjusted to pH 10 which might have caused precipitation of uranium phases.

The Np concentrations in the NaCl-rich solution are below the detection limit of 2×10^{-13} mol/l and in the $MgCl_2$-rich solution Np concentrations were only slightly above. The neptunium concentration in both solutions is determined by sorption onto the corroded cement phases. The sorption of Np (V) to cement paste in brines has been previously studied [103] and measured sorption coefficients were in the range of 1370 ml/g. However, in these experiments a lower cement mass to solution ratio was chosen and the pH_{exp} was 6.7. One can assume that the hardened cement paste have been degraded completely (dissolution of all C-S-H phases) and the sorption coefficient refers to secondary phases only. Such a high distribution coefficient may explain the low measured Np concentrations and the homogeneous distribution of Np in the cementitious matrix.

The calculated geochemical evolution of the solutions as function of the cement mass to solution volume shows an excellent agreement with the measured values of the long-term leaching and corrosion tests with simulated cemented waste full-scale blocks.

A temperature effect on the Ca - Mg exchange reaction has not been demonstrated experimentally in the temperature range from 28°C to 40°C. However, the mobilization of Cs was significantly faster for cemented full-scale blocks stored at 40°C in comparison to 28°C. Real radioactive waste forms, prepared at W/C ratio of 0.63 l/kg, showed even faster release: After 4.5 years already 90% of the cesium inventory was released. The same release was observed only after 15 years from blocks of W/C ratios of 0.43 l/kg.

Interactions of $MgCl_2$-rich solution with high quality concrete materials were not part of this project. It was referred to some data from other projects in chapter 1.2.3. The results indicate that the major interaction mechanisms between the solution components and concrete are the same. Within 11 years, the penetration of chloride was observed down to 12 cm into the concrete. The Ca-Mg-exchange processes were identified down to a depth of 5 cm in the concrete.

References

[1] M. Jacob, "Statistische Auswertung der im Jahre 1978 von der HDB an das Versuchsendlager Asse II abgelieferten radioaktiven Abfälle," 1981.

[2] Bergamt Goslar, "Endlagerung schwachradioaktiver Abfälle im Salzbergwerk Asse," *Genehmigung,* vol. 4777/75 II vom 29.12.1975, 1975.

[3] R. Köster, "Zusammenstellung der vom Arbeitskreis erarbeiteten Vorschläge zur Charakterisierung, Qualitätssicherung und Datensammlung von LAW/MAW-Produkten," INE, Karlsruhe, 1982.

[4] E. R. Merz, D. Dyckerhoff, and R. Odoj, " Characterization of radioactive wastes incorporated in a cement matrix," *Nuclear Journal of Canada,* vol. 1, pp. 173-178, 1987.

[5] P. Vejmelka, G. Rudolph, W. Kluger, and R. Köster, "Die Konditionierung radioaktiver Abfallösungen durch Zementierung," Forschungszentrum Karlsruhe, KfK 4800, 1990.

[6] Verein deutscher Zementwerke, *Zement Taschenbuch*, 50. Ausgabe ed. Wiesbaden - Berlin: Bauverlag, 2002.

[7] B. Kienzler and A. Loida, "Endlagerrelevante Eigenschaften von hochradioaktiven Abfallprodukten. Charakterisierung und Bewertung. Empfehlung des Arbeitskreises HAW-Produkte," Forschungszentrum Karlsruhe, FZKA-6651, 2001.

[8] G. G. Loomis, A. P. Zdinak, and M. T. Butte, "Field-scale permeation testing of jet-grouted buried waste sites. ." in *Waste Management '97, 2-7 Mar 1997*, Tucson, AZ, USA, 1997.

[9] BNL, "Lysimeter study of commercial reactor waste forms: waste form acquisition, characterization and full-scale leaching," Brookhaven National Lab, Report BNL--51613, 1983.

[10] J. Marcaillou, J. C. Faure, A. Bernard, and J. C. Nomine, "Practice followed at the Cadarache Nuclear Research Centre in the conditioning of solid radioactive wastes for purposes of temporary storage while awaiting disposal," in *International symposium on the conditioning of radioactive wastes for storage and disposal*, Utrecht (NL), 21-25 Jun 1982, 1982.

[11] D. C. Philips, R. Köster, and G. de Angelis, "Radiation, thermal and mechanical effects on low and medium active conditioned waste," in *Radioactive Waste Management and Disposal Conference*, Luxemburg, 1985.

[12] A. Bernard, A. Bonnet, G. Cornec, L. Farges, and J. C. Nomine, "Long-term leaching tests on full-scale blocks of radioactive wastes," *Nuclear and Chemical Waste Management,* vol. 3(3), pp. 161-168, 1982.

[13] M. Kikuchi, M. Matsuda, T. Nishi, S. Tamata, and T. Izumida, "A high performance cement for solidification of spent ion-exchange resin," in *'93 International conference on nuclear waste management and environmental remediation.*, Prague (Czech Republic) 5-11 Sept., 1993, pp. 177-181.

[14] J. C. Nomine, A. Billon, and G. Courtois, "Leaching scale effect for cement-waste forms," in *Mat. Res. Soc. Symp. Proc.*, 1989, pp. 431-438.

[15] P. Vejmelka, G. Rudolph, and W. Kluger, "Investigations of low and medium-level cemented waste forms from reprocessing Task 3 Characterization of radioactive waste forms A series of final reports (1985-89), No. 4," Nuclear Science and Technology, EUR 13544, 1991.

[16] W. Kluger, "persönliche Aufzeichnungen," 1980.

[17] M. I. Ojovan, G. A. Varlackova, Z. I. Golubeva, and O. N. Burlaka, "Long-term field and laboratory leaching tests of cemented radioactive wastes," *Journal of Hazardous Materials,* vol. 187, pp. 296-302, 2011.

[18] H. Hinsch and F. Kessler, "Leaching experiments with real-scale specimens in the Asse abandoned shaft. Period covered: 1979-1985," Report GSF--1/85, 1985.

[19] H. Hinsch, "Zwischenbericht über die Auslaugversuche in der Schachtanlage Asse," GSF, 01.11.1989, Nr. 04458, 1989.

[20] H. Hinsch, "Auslaugversuche 1990," GSF, 24.10.1990, 1990.

[21] H. Hinsch and F. Kessler, "Zwischenbericht über Auslaugversuche an Großproben in der Schachtanlage Asse," GSF, Institut für Tieflagerung, 1983.

[22] H. Hinsch and F. Kessler, "Auslaugversuche an Großproben in der Schachtanlage Asse, Versuchszeitraum 1979-1985,," GSF, Institut für Tieflagerung, Abteilung für Endlagertechnologie, 1985.

[23] B. Kienzler, P. Vejmelka, H. J. Herbert, H. Meyer, and C. Altenhein-Haese, "Lessons learnt from long-term leaching experiments of Full-scale cemented waste forms," in *Proc. Jahrestagung Kerntechnik'99*, 1999, pp. 351-354.

[24] B. Kienzler, P. Vejmelka, H.-J. Herbert, H. Meyer, and C. Altenhein-Haese, "Long-term leaching experiments fo full-scale cemented waste forms: Experiments and modeling," *Nuclear Technology,* vol. 129, pp. 101-118, 2000.

[25] B. Kienzler, M. Schlieker, W. Schüßler, V. Metz, D. Hentschel, C. Nies, N. Kerner, A. Seither, H. Meyer, and G. Bracke, "Langzeit Auslaug- und Korrosionsexperimente an zementierten 1:1 Gebinden in der Schachtanlage Asse: Probennahme und Auswertung 2001," Forschungszentrum Karlsruhe, FZKA 6716, 2002.

[26] B. Kienzler, M. Schlieker, A. Bauer, Volker Metz, and H. Meyer, "Langzeit Auslaug- und Korrosionsexperimente an zementierten 1:1 Gebinden in der Schachtanlage Asse: Probennahme und Auswertung 2003," Forschungszentrum Karlsruhe, FZKA 7059, 2004.

[27] B. Kienzler, "Zwischenbericht zur Phase II Analysen der Auslauglösungen beim FZK-INE und Gewinnung von Feststoff-Proben am Forschungszentrum Jülich (FZJ)," FZK-INE, Karlsruhe, FZK-INE 004/07, Analyse der Eigenschaften von simulierten, zementierten 1:1 Gebinden und ihres Phasenbestandes nach 20 Jahren Auslaugung in Salzlösungen, 2007.

[28] B. Kienzler, M. Schlieker, N. Müller, E. G. Robles, C. Borkel, V. Krepper, M. Plaschke, F. Geyer, S. Hilpp, M. Lagos, S. Moisei-Rabung, A. Seither, and C. Walschburger, "Experimentelles Programm zur Untersuchung von simulierten, zementierten 1:1 Gebinden aus dem Auslaugversuchsfeld Nr. 1 der Schachtanlage Asse: Teil II: Probenahme und Analyse der Lösungen sowie Präparation der Feststoffproben," INE, KIT-INE 001/2014, 2014.

[29] B. Kienzler, "Analyse der Eigenschaften von simulierten, zementierten 1:1 Gebinden und ihres Phasenbestandes nach 20 Jahren Auslaugung in Salzlösungen: Abschlussbericht," FZK-INE, Karlsruhe, FZK-INE 001/09, 2009.

[30] B. Kienzler, V. Metz, B. Brendebach, N. Finck, M. Plaschke, T. Rabung, J. Rothe, and D. Schild, "Chemical status of U(VI) in cemented waste forms under saline conditions," *Radiochim. Acta,* vol. 98, pp. 675-684, 2010.

[31] C. Borkel, B. Kienzler, M. Schlieker, V. Metz, D. Schild, E. Soballa, M. Lagos, C. Walschburger, S. Heck, S. Hilpp, S. Moisei-Rabung, and A. Seither, "Experimentelles Programm zur Untersuchung von simulierten, zementierten 1:1 Gebinden aus dem Auslaugversuchsfeld Nr. 1 der Schachtanlage Asse: Teil III Ergebnisse der Festkörperanalysen " Institut für Nukleare Entsorgung (INE), Karlsruhe Institute of Technology (KIT), 2015.

[32] B. Kienzler and V. Metz, "Cemented waste forms: 25 years full-scale leaching experiments.," in *10th Internat. Conf.on Environmental Remediation and Radioactive Waste Management*, Glasgow, UK, September 4-8, 2005.

[33] J. Rothe, B. Brendebach, C. Bube, K. Dardenne, M. A. Denecke, B. Kienzler, V. Metz, T. Prüßmann, K. Rickers-Appel, D. Schild, E. Soballa, and T. Vitova, "Characterization of U(VI)-phases in corroded cement products by micro(μ)-spectroscopic methods," *Journal of Physics: Conference Series,* vol. 430 p. 012114, 2013.

[34] B. Kienzler, V. Metz, and A. Bauer, "Interactions of cemented waste forms with salt brines: Results after 25 years full-scale tests," in *5th International Seminar on Radioactive Waste Products (RADWAP 2008)*, Würzburg, Germany, 27 - 31 October, 2008.

[35] C. Bube, V. Metz, D. Schild, J. Rothe, K. Dardenne, M. Lagos, M. Plaschke, and B. Kienzler, "Combining thermodynamic simulations, element and surface analytics to study U(VI) retention in corroded cement monoliths upon > 20 years of leaching," *Physics and Chemistry of the Earth,* vol. 70-71, pp. 53-59, 2014.

[36] C. Bube, V. Metz, D. Schild, M. Lagos, E. Bohnert, K. Garbev, M. Altmaier, and B. Kienzler, "Interactions of U(VI) with cement alteration products in highly saline solutions," presented at the 1st

Internat. Symp. on Cement-based Materials for Nuclear Wastes (NUWCEM 2011), Avignon, F, October 11-14, 2011 2011.
[37] B. Kienzler, V. Metz, J. Lützenkirchen, E. Korthaus, and T. Fanghänel, "Geochemically based safety assessment," *Journal of Nuclear Science and Technology,* vol. 44, pp. 470-476, 2007.
[38] B. Kienzler, M. Altmaier, C. Bube, and V. Metz, "Radionuclide Source Term for Irradiated Fuel from Prototype, Research and Education Reactors, for Waste Forms with Negligible Heat Generation and for Uranium Tails ", KIT Scientific Reports 7635, Vorläufige Sicherheitsanalyse Gorleben (vSG), 2013.
[39] W. Kluger, R. Götz, and F. Dirks, "Experimentelle Untersuchungen zur Barrferenwirkung von Blechfässern und verlorenen Betonabschirmungen im Hinblick auf die Auslaugung von 200 1-Zementgebinden in quinärer Lauge: 1. Zwischenbericht," Institut für Nukleare Entsorgungstechnik (INE), Forschungszentrum Karlsruhe, 1980.
[40] H. U. Bambauer, "Mineralogische und chemische Untersuchunge zur Langzeitwirkung einer Salzlauge auf einen Betonmantel," Westfälische Wilhelms-Universität Münster, Institut für Mineralogie Münster, 1992.
[41] P. Vejmelka, R. Köster, D. Ferrara, and M. E. Wacks, "Leach studies of chelating agents and influence on radionuclide leaching from simulated IIW/ILW cement waste forms," in *Waste Management'90,* Tucson, February 25 - March 1, 1990, pp. 49-56.
[42] C. Altenhain-Haese, H. Bischoff, and G. Marx, "Quellterme für HAW-Glas, abgebrannten Kernbrennstoff und zementierte Abfälle: Auswertung der Langzeit-Auslaugexperimente an 1:1 Gebinden im Forschungsbergwerk Asse," Institut für Nukleare Entsorgung (INE), Forschungszentrum Karlsruhe Karlsruhe, FZK-INE 008/99, Erstellung eines integrierten Nahfeldmodells von Gebinden hochaktiver Abfälle im Salzstock Gorleben: geochemisch fundierter Quellterm für HAW-Glas, abgebrannte Brennelemente und Zement, 1999.
[43] R. Köster, "Cementation of Radioactive Waste in the Federal Republic of Germany," in *Waste management '85,* Tucson, AZ (USA), 24-28 Mar. 1985, 1985, pp. 487-491.
[44] H. Hinsch and F. Keßler, "Auslaugversuche an Großproben in der Schachtanlage Asse," Gesellschaft für Strahlen- und Umweltforschung, München, GSF-Bericht 1/85, 1985.
[45] K. Siemon and P. Patzelt, "Title," unpublished|.
[46] T. Fanghänel, V. Neck, and J. I. Kim, "The Ion Product of H_2O, Dissociation Constants of H_2CO_3 and Pitzer Parameters in the System $Na^+/H^+/OH^-/HCO_3^-/CO_3^{2-}/ClO_4^-/H_2O$ at 25°C," *Journal of Solution Chemistry,* vol. 25, pp. 327-343, 1996.
[47] A. R. Felmy and J. H. Weare, "Calculation of multicomponent ionic diffusion from zero to high concentration: I. The system Na-K-Ca-Mg-Cl-SO4-H2O at 25°C," *Geochimica et Cosmochimica Acta,* vol. 55, pp. 113-131, 1991/01/01 1991.
[48] M. Altmaier, V. Metz, V. Neck, R. Müller, and T. Fanghänel, "Solid-liquid equilibria of $Mg(OH)_2$(cr) and $Mg_2(OH)_3Cl_4H_2O$(cr) in the system Mg-Na-H-OH-Cl-H_2O at 25°C," *Geochimica et Cosmochimica Acta,* vol. 67, pp. 3595-3601, 2003.
[49] M. Altmaier, V. Neck, and T. Fanghanel, "Solubility of Zr(IV), Th(IV) and Pu(IV) hydrous oxides in CaCl2 solutions and the formation of ternary Ca-M(IV)-OH complexes," *Radiochimica Acta,* vol. 96, pp. 541-550, 2008.
[50] B. Kienzler, A. Bauer, and M. Schlieker, "Zwischenbericht zur Phase I Vorbereitende Arbeiten und Transport der Gebinde von der Schachtanlage Asse nach dem Forschungszentrum Jülich (FZJ) bzw. nach dem Forschungszentrum Karlsruhe (FZK)," FZK-INE, Karlsruhe, FZK-INE 00x/07, Analyse der Eigenschaften von simulierte, zementierten 1:1 Gebinden und ihres Phasenbestandes nach 20 Jahren Auslaugung in Salzlösungen, 2007.
[51] T. Rabung and H. Geckeis, "Influence of pH and metal ion loading on the Cm(III) humate complexation: a time resolved laser fluorescence spectroscopy study," *Radiochimica Acta,* vol. 97, pp. 265-271, 2009.
[52] A. Denecke, J. Rothe, K. Dardenne, H. Blank, and J. Hormes, "The INE-beamline for actinide research at ANKA," *Physika Scripta,* vol. T115, pp. 1001-1003, 2005.
[53] W. Riedel, "Die Korrosionsbeständigkeit von Zementmörteln in Magnesiumsalzlösungen," *Zement, Kalk, Gips,* vol. 6, pp. 286-296, 1973.

[54] H. G. Smolczyk, "Some observations and new aspects concerning sea-water action on concrete in the tidal zone," *Rilem Bulletin, 1966/1967,* vol. 30, 1967.

[55] J. Heppner, S. Peter, and O. Weickhard, "Zur Beständigkeit von Tiefbohrzementen gegenüber Formationswässern mit höherem Gehalt an Magnesium," *Erdöl-Erdgas-Zeitschrift,* vol. 84, p. 371, 1968.

[56] J. Bonzel and F. W. Locher, "Über das Angriffsvermögen von Wässern, Böden und Gasen auf Beton: Anmerkungen zu den Normentwürfen DIN 4030 E und DIN 1045 E," https://www.vdz-online.de/fileadmin/gruppen/vdz/3LiteraturRecherche/Veroeffentlichungeninfachzeitschriften/PDF/24628.pdf.

[57] H.-J. Herbert, "Auslaugversuche in der Schachtanlage Asse, Zwischenbericht 01.11.89-31.05.92," Institut für Tieflagerung, Abteilung für Endlagersicherheit., GSF, Abteilungsbericht IfT 4/92, 1992.

[58] B. Kienzler, M. Schlieker, N. Müller, V. Metz, N. Finck, and M. Lagos, "Auslaugversuchsfeld Asse: Zwischenbericht Ergebnisse der Lösungsanalysen 2011," INE, KIT-INE 004/2012, 2012.

[59] V. G. Petrov, S. N. Kalmykov, and M. Altmaier, "Solubility and Phase Transformations of Np(V) Hydroxide in Solutions with Different Ionic Strengths," *Moscow University Chemistry Bulletin,* vol. 66, pp. 107–115, 2011.

[60] W. Erfurt, "Erfassung von Gefügeveränderungen in Beton durch Anwendung zerstörungsfreier Prüfverfahren zur Einschätzung der Dauerhaftigkeit," Dr. ing., F. A. Finger-Institut für Baustoffkunde, Bauhaus-Universität Weimar, 2002.

[61] Verein deutscher Zementwerke, *Zement Taschenbuch*, 48. Ausgabe ed. Wiesbaden - Berlin: Bauverlag, 1984.

[62] C. Borkel, V. Montoya, and B. Kienzler, "Modeling long-term leaching experiments of full scale cemented wastes: effect of solution composition on diffusion," in *NUWCEM 2014, 2nd International Symposium on Cement-based Materials for Nuclear Wastes*, Avignon, 03-06 June 2014, 2014.

[63] L. G. Baquerizo, T. Matschei, K. L. Scrivener, M. Saeidpour, A. Thorell, and L. Wadsö, "Methods to determine hydration states of minerals and cement hydrates," *Cement and Concrete Research,* vol. 65, pp. 85-95, 11// 2014.

[64] N. Macé, E. Wieland, R. Dähn, J. Tits, A. C. Scheinost, T. Stumpf, and C. Walther, "Determination of the local coordination environment of U(VI) sorption species on cementitious materials under alkaline conditions," in *MRS'09, Scientific Basis for Nuclear Waste Management*, St Peterburg, Russia, 2009.

[65] E. Wieland, N. Macé, R. Dähn, D. Kunz, and J. Tits, "Macro- and micro-scale studies on U(VI) immobilization in hardened cement paste," *Journal of Radioanalytical and Nuclear Chemistry,* vol. 286, pp. 793-800, 2010.

[66] G. Geipel, T. Reich, V. Brendler, G. Bernhard, and H. Nitsche, "Laser and X-ray spectroscopic studies of uranium-calcite interface phenomena," *Journal of Nuclear Materials,* vol. 248, pp. 408-411, 1997.

[67] G. Geipel, G. Bernhard, M. Rutsch, V. Brendler, and H. Nitsche, "Spectroscopic properties of uranium(VI) minerals studied by time-resolved laser-induced fluorescence spectroscopy (TRLFS)," *Radiochimica Acta,* vol. 88, pp. 757-762, 2000.

[68] G. Bernhard, "Speciation of uranium in environmental relevant compartments," *Landbauforschung Volkenrode,* vol. 55, pp. 139-148, 2005.

[69] D. Ginderow, "Structure de l'uranophane alpha, $Ca(UO_2)_2(SiO_3OH)\bullet 2.5H_2O$," *Acta Crystallographica,* vol. 44, Section C, pp. 421-424, 1988.

[70] C. Bube, V. Metz, E. Bohnert, K. Garbev, D. Schild, and B. Kienzler, "Long-term cement corrosion in chloride-rich solutions relevant to radioactive waste disposal in rock salt – Leaching experiments and thermodynamic simulations," *Physics and Chemistry of the Earth,* vol. 64, pp. 87–94, 2013.

[71] I. Grenthe, J. Fuger, R. J. M. Konings, R. J. Lemire, A. B. Muller, C. Nguyen-Trung, and H. Wanner, *Chemical Thermodynamics of Uranium*. Amsterdam, The Netherlands: Nuclear Energy Agency, Organisation for Economic Co-operation, North Holland Elsevier Science Publishers, 1992

[72] R. Guillaumont, T. Fanghänel, J. Fuger, I. Grenthe, V. Neck, D. Palmer, and M. Rand, *Update on the cemical thermodynamics of uranium, neptunium, plutonium, americium, and technetium, NEA Organization f Cooperation and Development* vol. 5. Issy-les-Moulineaux, France: NEA, 2003.

[73] V. Neck and M. Altmaier, "Solubility of U(VI) in NaCl and MgCl2 solutions," in *Migration`03*, Gyeongju (Korea), 2003.
[74] B. Kienzler, M. Altmaier, C. Bube, and V. Metz, "Radionuclide Source Term for HLW Glass, Spent Nuclear Fuel, and Compacted Hulls and End Pieces (CSD-C Waste)," KIT-INE 003/11, KIT Scientific Reports 7624, Vorläufige Sicherheitsanalyse Gorleben (VSG), 2012.
[75] R. J. Lemire, *Chemical thermodynamics of neptunium and plutonium* vol. 4: Elsevier, 2001.
[76] V. Neck, M. Altmaier, R. Müller, V. Metz, and B. Kienzler, "Teil 3: Löslichkeitsexperimente zur Absicherung der thermodynamischen Datenbasis," FZK-INE, Karlsruhe, FZK-INE 007/03, Experimentelles Programm zur Bestätigung der Ergebnisse von standortspezifischen Modellrechnungen für die Schachtanlage Asse, 2003.
[77] V. Petrov, X. Gaona, D. Fellhauer, J. Rothe, K. Dardenne, S. Kalmykov, and M. Altmaier, "Migration Book of Abstracts," in *Migration*, UK, 2013.
[78] D. Fellhauer, "laufendes Projekt an KIT-INE, private Mitteilung," 2015.
[79] W. Runde, M. P. Neu, and D. L. Clark, "Neptunium(V) hydrolysis and carbonate complexation: Experimental and predicted neptunyl solubility in concentrated NaCl using the Pitzer approach," *Geochimica et Cosmochimica Acta,* vol. 60, pp. 2065-2073, 6// 1996.
[80] C. Bethke and S. Yeakel, "The Geochemist's Workbench," University of Illinois, Urbana, USA 2009.
[81] K. S. Pitzer, "Theory: Ion interaction Approach," in *Activity coefficients in electrolyte solutions*. vol. 7, R. M. Pytkowicz, Ed., ed: CRC Press, INC. Boca Raton Fl., 1979, pp. 157-208.
[82] C. E. Harvie, N. Møller, and J. H. Weare, "The prediction of mineral solubilities in natural waters: The Na-K-Mg-Ca-H-Cl-SO4-OH-HCO3-CO3-CO2-H2O system to high ionic strengths at 25°C," *Geochimica et Cosmochimica Acta,* vol. 48, pp. 723-751, 1984.
[83] E. J. Reardon, "Problems and approaches to the prediction of the chemical composition in cement/water systems," *Waste Manag,* vol. 12, pp. 221-239, 1992.
[84] W. Hummel, U. Berner, E. Curti, F. J. Pearson, and T. Thoenen, "Nagra/PSI Chemical Thermodynamic Data Base 01/01," Nagra, Wettingen, Switzerland, 1015-2636, Technical Report, 589, 2002.
[85] E. J. Reardon, "Ion interaction parameters for aluminum sulfate and application to the prediction of metal sulfate solubility in binary salt systems," *The Journal of Physical Chemistry,* vol. 92, pp. 6426-6431, 1988/11/01 1988.
[86] E. J. Reardon, "An ion interaction model for the determination of chemical equilibria in cement/water systems," *Cement and Concrete Research,* vol. 20, pp. 175-192, 1990.
[87] R. A. Robie and B. S. Hemingway, "Thermodynamic Properties of Minerals and Related Substances at 298. 15 K and 1 bar (105 Pascals) Pressure and at Higher Temperatures," 1995.
[88] B. Lothenbach and F. Winnefeld, "Thermodynamic modelling of the hydration of Portland cement," *Cement and Concrete Research,* vol. 36, pp. 209-226, 2006.
[89] T. Matschei, B. Lothenbach, and F. P. Glasser, "Thermodynamic properties of Portland cement hydrates in the system CaO–Al2O3–SiO2–CaSO4–CaCO3–H2O," *Cement and Concrete Research,* vol. 37, pp. 1379-1410, 2007.
[90] S. A. Stronach and F. P. Glasser, "Modelling the impact of abundant geochemical components on phase stability and solubility of the CaO–SiO2–H2O system at 25°C: Na^+, K^+, SO_4^{2-}, Cl^- and CO_3^{2-}," *Advances in Cement Research,* vol. 9, pp. 167-181, 1997.
[91] T. J. Wolery, "EQ3/6, A Software Package for Geochemical Modeling of Aqueous Systems," University of California, Lawrence Livermore National Laboratory 1992.
[92] R. Guillaumont, T. Fanghänel, J. Fuger, I. Grenthe, V. Neck, D. A. Palmer, M. H. Rand, F. J. Mompean, M. Illemassene, C. Domenech-Orti, and K. Ben-Said, *Update on the Chemical Thermodynamics of Uranium, Neptunium, Plutonium, Americium and Technetium*: Elsevier Science, 2003.
[93] M. Altmaier, V. Neck, R. Mueller, and T. Fanghänel, "Solubility of U(VI) and formation of $CaU_2O_7x3H_2O$(cr) in alkaline $CaCl_2$ solutions," presented at the MIGRATION 2005, 10th International Conference on Chemistry and Migration Behaviour of Actinides and Fission Products in the Geosphere, Avignon (France); 18-23 Sep 2005, 2005.
[94] V. Neck, M. Altmaier, R. Mueller, M. Schlieker, and T. Fanghänel, "Solubility of U(VI) in NaCl and $MgCl_2$ solutions," presented at the MIGRATION 2003, 9th International Conference on Chemistry and Migration Behaviour of Actinides and Fission Products in the Geosphere, Gyeongju (Korea), 2003.

[95] K. S. Pitzer, *Activity coeeficients in electrolyte solutions*, 2. ed. Boca Raton, Fl., USA: CRC Press, Inc., 1991.

[96] A. Plyasunov, T. Fanghänel, and I. Grenthe, "Estimation of the Pitzer Equation Parameters for Aqueous Complexes. A Case Study for Uranium at 298.15 K and 1 atm," *Acta Chemica Scandinavica,* vol. 52, 1998.

[97] I. Grenthe and A. Plyasunov, "On the use of semiempirical electrolyte theories for modeling of solution chemical data. , ," *Pure and Applied Chemistry,* vol. 69, pp. 951-958, 1997.

[98] A. Plyasunov, T. Fanghänel, and I. Grenthe, "Estimation of the Pitzer equation parameters for aqueous complexes. A case study for uranium at 298.15 K and 1 atm," *Acta Chemica Scandinavica,* vol. 52, pp. 250–260, 1998.

[99] E. J. Reardon, "Problems and Approaches to the Prediction of the Chemical Composition in Cement/Water Systems," *Waste Management,* vol. 12, pp. 221-239, 1992.

[100] E. J. Reardon, "An ion interaction model for the determination of chemical equilibria in cement/water systems," *Cement and Concrete Research,* vol. 20, pp. 175-192, 1990.

[101] B. Kienzler and V. Metz, "Modelling long-term corrosion of cemented waste forms in salt brines," in *12th Internat. Conf. on Environmental Remediation and Radioactive Waste Management (ICEM '09)*, Liverpool, GB, October 11-15, 2009.

[102] A. Jakob, F. A. Sarott, and P. Spieler, "Diffusion and sorption on hardened cement pastes - experiments and modelling results," PSI Report No. 99-05, 1999.

[103] J. Lützenkirchen, P. Vejmelka, V. Metz, and B. Kienzler, "Site specific sorption data for the Asse salt mine," in *The 9th International Conference on Radioactive Waste Management and Environmental Remediation ICEM '03*, Oxford, England, September 21 – 25, 2003.

Acknowledgement

The project described in this report was funded by the Federal Ministry of Education and Research (Bundesministerium für Bildung und Forschung) under contract 02ALVF2012. The sole responsibility for the report's contents lies with the authors.

The authors thank the staff of the Asse II salt mine as well as the Asse-GmbH for supervision of the experimental room and their support during the sampling campaigns. Our thanks and appreciation to the employees of companies and institutions for excellent cooperation and their readiness to support special wishes with respect to fixation of the full-scale samples and the transports. For preparation of drilling cores and sampling of drilling dust the authors thank the staff of FZJ-IEK-6 and of HDB at WAK GmbH.

Our thanks to the employees of the Institut für Funktionelle Grenzflächen of KIT for some thermogravimetric analyses.

The investigations would not have been possible without financial support by the Federal Office of Radiation Protection (BfS) and Helmholtz-Zentrum München (HMGU).

The authors would like to thank the staff of the Institut für Nukleare Entsorgung (INE) for their excellent work and for their high degree of commitment.